SpringerBriefs in Computer Science

Series Editors

Stan Zdonik
Peng Ning
Shashi Shekhar
Jonathan Katz
Xindong Wu
Lakhmi C. Jain
David Padua
Xuemin Shen
Borko Furht
V. S. Subrahmanian
Martial Hebert
Katsushi Ikeuchi
Bruno Siciliano

For further volumes:
http://www.springer.com/series/10028

Austin Parker • Gerardo I. Simari
Amy Sliva • V.S. Subrahmanian

Data-driven Generation of Policies

 Springer

Austin Parker
Department of Computer Science
University of Maryland
College Park, MD, USA

Gerardo I. Simari
Department of Computer Science
University of Oxford
Oxford, UK

Amy Sliva
Charles River Analytics Inc.
Cambridge, MA, USA

V.S. Subrahmanian
Computer Science Department
University of Maryland
College Park, MD, USA

ISSN 2191-5768   ISSN 2191-5776 (electronic)
ISBN 978-1-4939-0273-6   ISBN 978-1-4939-0274-3 (eBook)
DOI 10.1007/978-1-4939-0274-3
Springer New York Heidelberg Dordrecht London

Library of Congress Control Number: 2013956513

Printed on acid-free paper

Springer is part of Springer Science+Business Media (www.springer.com)

Acknowledgements

The majority of this work was carried out when the authors were at the University of Maryland, College Park. Some of the authors of this book may have been funded in part by AFOSR grant FA95500610405, ARO grants W911NF0910206, W911NF1160215, W911NF1110344, an ARO/Penn State MURI award, and ONR grant N000140910685. Gerardo Simari is also currently partially supported by the Engineering and Physical Sciences Research Council of the United Kingdom (EPSRC) grant EP/J008346/1 ("PrOQAW: Probabilistic Ontological Query Answering on the Web"), by European Research Council grant 246858 ("DIADEM: Domain-centric Intelligent Automated Data Extraction Methodology"), and by a Google Research Award.

Author Bios

Austin Parker:

Austin Parker got his Ph.D. from the University of Maryland in 2008 with a thesis entitled "Spatial Temporal Probabilistic Databases". In graduate school he also did work on AI game playing, writing an algorithm to play an imperfect-information chess variant called Kriegspiel, which won the second place in the 2006 Computer Chess Olympics. In his undergraduate work at Haverford College, he double majored in Mathematics and Philosophy, writing programs to test the validity of important conjectures, and to model interesting biological processes. He currently lives in Maryland.

Gerardo I. Simari:

After completing the *Licenciatura* program in Computer Science at Universidad Nacional del Sur (Bahia Blanca, Argentina), Gerardo Simari began his research career in 2003, when he was awarded a prestigious fellowship by the *Comisión de Investigaciones Científicas de la Provincia de Buenos Aires* (Argentina) to fund his studies towards a Master's degree in Computer Science under Simon Parsons (Brooklyn College, City University of New York, USA) as a member of the Artificial Intelligence Research and Development Laboratory (LIDIA) at Universidad Nacional del Sur. In 2005, he started his Ph.D. studies at University of Maryland College Park (USA) under V. S. Subrahmanian. During his time in Maryland, he was a central participant in several research projects carrying out work in probabilistic logic programming, with applications to stochastic reasoning about behaviors of groups of interest, and reasoning in the presence of large amounts of inconsistent data. He has published more than 40 papers in high-quality international conferences and journals, served as reviewer for top-tier international conferences and journals, and was awarded the Best Student Paper prize (shared with Matthias

Bröcheler) at the 2009 International Conference on Logic Programming (ICLP). He is currently a senior researcher at the Department of Computer Science, University of Oxford (UK).

Amy Sliva:

Amy Sliva is a Scientist at Charles River Analytics researching artificial intelligence models and large-scale data analytics for decision-making. Dr. Sliva was previously an Assistant Professor of Computer Science and Political Science at Northeastern University where she developed interdisciplinary tools for understanding, forecasting, and responding to behavioral dynamics in international conflict, security policy, and international development. She has collaborated with the National Defense University on analysis of the strategic aspects of cyber warfare and worked for the World Bank developing similar behavioral modeling technologies for education development in Nigeria. Dr. Sliva received her Ph.D. in Computer Science from the University of Maryland in 2011, where she worked in the Laboratory for Computational Cultural Dynamics to create decision-support tools for the National Security and Intelligence Communities for counterterrorism analysis. She also has a B.S. in Computer Science from Georgetown University (2005), an M.S. in Computer Science from the University of Maryland (2007), and a Master of Public Policy (M.P.P.) in International Security and Economic Policy from the University of Maryland (2010).

V. S. Subrahmanian:

V. S. Subrahmanian is a Professor of Computer Science and Director of the Center for Digital International Government and Co-Director of the Laboratory for Computational Cultural Dynamics at the University of Maryland, where he has been on the faculty since 1989. He has served on the editorial board of several journals, has won numerous awards, and delivered invited talks at numerous conferences. He has worked extensively on databases and artificial intelligence and has co-authored over 200 papers as well as several books.

Contents

Chapter 1
Introduction and Related Work

A large number of well known data sets in the social sciences have a tabular form. Each row refers to a period of time, and each column represents a variable that characterizes the state of some entity during a time period. These variables naturally divide into those actionable variables we can control (which we will call "action variables") and those we cannot (which we will call "state variables"). For example, data sets regarding school performance for various U.S. states contain "state variables" such as the graduation rate of students in the state and the student to staff ratio during some time frame, while the "action" variables might refer to the level of funding provided per student during that time frame, the faculty salary levels during that time period, etc. Clearly, policy makers U.S. state can attempt to change the levels of funding per student and/or change the faculty salaries in an attempt to increase the graduation rate. In a completely different setting, political science data sets about the stability of a country (such as the data sets created by the well known Political Instability Task Force [4]) may have "state variables" such as the GDP of a country during a time period, the infant mortality rate during the same time period and the number of people killed in political conflict in the country during that time period, while "action" variables might include information about the investment in hospitals or education during that time frame, the number of social workers available, and so forth. A government might want to see what actionable policies it can try to implement to achieve a certain goal (e.g., bringing the infant mortality rate below some threshold).

These are just two examples of *problems that are not easily solved using current algorithms for reasoning about actions in AI or by AI planning systems*. The main reasons are the following:

1. The relationships between the actions and their impact on the state are poorly understood;
2. A set of actions, taken together, might have a cumulative effect on a state that might somehow be more than a naive combination of the effects of those actions individually—which of course are not known anyway; and

A. Parker et al., *Data-driven Generation of Policies*, SpringerBriefs in Computer Science, DOI 10.1007/978-1-4939-0274-3_1, © The Author(s) 2014

3. The actions under consideration may not succeed—an attempt to raise hospital funding may be blocked for reasons outside of anyone's control. In essence, little reliable quantitative information exists on the correlative aspects of multiple actions and whether these are somehow conditioned on other variables.

In this work, we first propose (Sect. 1.1) the notion of an *event KB* (this is not novel, but generalizes several social science data sets such as those mentioned above). Chapter 2 defines the concept of "state change attempts" (SCAs for short) representing policies devised with the objective of reaching a certain goal, and formulates various problems related to finding "optimal" (in a sense we will make precise) SCAs relative to a goal. We develop a host of results on the computational complexity of finding optimal SCAs. In Chap. 3, we study different kinds of effect estimators and arrive at one that is especially useful due to its computational properties. In Sect. 3.3, we first present a straightforward algorithm called **DSEE_OSCA** to compute optimal SCAs. We then develop a vastly improved algorithm called **TOSCA** based on tries. Though tries are a well known data structure, the novelty of our work is rooted in how **TOSCA** uses tries to solve optimal SCA computation problems with lower computational complexity. In Chap. 4, we discuss the formal relationship between computing solutions to OSCA problems and solving Markov Decision Processes (which we take as representatives of problems in the area of planning under uncertainty). Finally, in Chap. 5, we briefly describe an implementation of both algorithms, together with an experimental analysis (that uses both synthetic and real-world education data) to demonstrate that **TOSCA** is fast and can be effectively used on real-world data sets.

1.1   Preliminaries on Event KBs

An event KB is a relational database whose rows correspond to some time period (explicit or implicit) and whose columns are of two types—*state attributes* and *action attributes*. Throughout this work, we will assume the existence of some arbitrary, but fixed set $\mathbf{A} = \{A_1, \ldots, A_n\}$ of action attributes (also referred to as the "action schema"), and another arbitrary, but fixed set $\mathbf{S} = \{S_1, \ldots, S_m\}$ of state attributes (also referred to as the "state schema"). As usual, each attribute (state or action) A has a domain $dom(A)$, which in this work we assume to be finite. A *tuple* with respect to $(\mathbf{A}, \mathbf{S})$ is any member of $dom(A_1) \times \cdots \times dom(A_n) \times dom(S_1) \times \cdots \times dom(S_m)$. We use $t(S_i)$ (resp. $t(A_j)$) in the usual way to denote the value assigned to attribute S_i (resp. A_j) by a tuple. An *event knowledge base* $\mathcal{K}$ is a finite set of tuples with respect to $(\mathbf{A}, \mathbf{S})$. We assume all attributes A have domain $dom(A) \subset \mathbb{R}$. We use $\mathcal{A}$ to represent the set $dom(A_1) \times \cdots \times dom(A_n)$ and $\mathcal{S}$ to represent the set $dom(S_1) \times \cdots \times dom(S_m)$. We say a tuple is an *action tuple* if it contains only values for the action attributes and that it is a *state tuple* if it contains only values for the state attributes.

	A_1	A_2	A_3	A_4	S_1	S_2	S_3	S_4	S_5
t_1:	9,532	61.6	7.8	4.2	81.1	49.1	51.3	50.6	Yes
t_2:	9,691	63.2	7.8	5.7	82.3	52.1	54.6	53.3	No
t_3:	9,924	63.8	8.1	3.1	82.0	59.8	60.4	60.1	Yes
t_4:	10,148	64.2	7.6	3.4	83.4	60.5	64.2	63.3	Yes
t_5:	10,022	64.0	7.2	2.9	83.2	63.9	68.9	66.9	Yes

Fig. 1.1 Small instance of an event KB containing hypothetical school performance data

	A_1	A_2	A_3	A_4	A_5	A_6	A_7	A_8	A_9	A_{10}	A_{11}	A_{12}	S_1	S_2	S_3	S_4	S_5	S_6
t_1:	2500	5.0	6.5	5.5	350	120	70	450	430	12	150	950	45	21	130	15	41	12
t_1:	2610	5.5	6.0	5.0	380	180	80	470	450	15	150	950	48	18	123	18	46	11
t_2:	2750	5.0	6.0	5.5	360	180	80	410	490	21	150	960	42	25	145	21	35	9
t_3:	2800	5.0	5.5	6.0	350	180	90	500	560	20	155	970	51	31	160	12	32	9
t_4:	3200	6.0	5.5	6.0	350	190	90	510	565	25	155	1000	53	35	175	14	31	11
t_5:	3350	6.0	5.5	6.0	400	200	95	510	570	27	160	1005	55	24	152	15	62	8
t_6:	3400	6.0	5.5	5.5	450	200	100	535	565	30	160	1075	46	12	146	10	67	11
t_7:	3500	6.0	5.5	5.5	420	220	100	515	540	32	150	1100	41	15	135	16	59	6
t_8:	3550	6.0	5.5	5.5	420	210	100	570	600	32	120	1150	38	16	138	15	56	5
t_9:	3900	6.0	5.5	5.5	520	210	110	550	600	35	130	1200	39	9	98	8	85	7
t_{10}:	4010	6.0	5.5	5.5	550	215	120	550	605	40	135	1200	41	5	82	7	89	4

Fig. 1.2 Small instance of an event KB containing action and state data for a city government. Each tuple represents 1 year

Throughout this work, we will refer to two sample datasets, portions of which appear in Figs. 1.1 and 1.2. In the following examples we discuss these two datasets.

Example 1.1. In Fig. 1.1, we present a small portion of a *school* event KB containing data related to school performance in some region. The columns labeled $A_1, \ldots, A_4$ represent action attributes, i.e., variables that are subject to influence:

- A_1: Funding ($/Student),
- A_2: Salaries (% of Total Funding),
- A_3: Student/Staff Ratio, and
- A_4: Proficiency Increase Target.

while the columns labeled $S_1, \ldots, S_5$ represent state attributes, i.e., those that depend on the values of action attributes and are not subject to direct influence:

- S_1: Graduation (%),
- S_2: Math Proficiency,
- S_3: Reading Proficiency,
- S_4: Proficiency Score, and
- S_5: Target Reached (Y/N).

Math and reading scores obtained from standardized tests are combined into one annual *proficiency score*. School administrators have the goal of increasing proficiency and graduation percentages by certain amounts. The policies they create will involve influencing the action attributes towards this end.   ∎

Example 1.2. The second dataset we consider gives rise to what we will call a *city* event KB, a small example of which is given in Fig. 1.2. The KB contains information regarding taxes, funding, how funding is allocated, crime rates, arrests, and other aspects of interest for a city government. This kind of data has in the past decade slowly become available to the public through *open government* outlets such as www.data.gov, www.data.gov.uk, or the so-called *data blogs* such as those maintained by The New York Times[1] or The Guardian,[2] among many others harboring many different kinds of data.

In our simple example, action attributes are represented in the columns labeled $A_1, \ldots, A_{12}$:

- A_1: Overall city budget (in millions of dollars);
- A_2: Corporate tax rate (%): The domain of this attribute can be assumed to be an ordered set of values such as $dom(A_2) = \{0.0, 0.5, 1.0, 1.5, \ldots, 50.0\}$;
- A_3: Personal income tax (%): The domain can be assumed to be as in A_2;
- A_4: Sales tax: The domain can be assumed to be as in A_2;
- A_5: Funding for police department (in millions of dollars);
- A_6: Funding for fire department (in millions of dollars);
- A_7: Funding for street lighting (in thousands of dollars);
- A_8: Funding for traffic control (in thousands of dollars);
- A_9: Funding for public works: roads and bridges (in thousands of dollars);
- A_{10}: Funding for city hospital (in millions of dollars);
- A_{11}: Funding for parks and recreation (in thousands of dollars); and
- A_{12}: Funding for education (in millions of dollars).

The state attributes are given by the columns labeled $S_1, \ldots, S_6$, and can be described as follows:

- S_1: Government approval rating (% of the population);
- S_2: Fire-related incidents involving serious property damage;
- S_3: Petty crimes;
- S_4: Felonies;
- S_5: Arrests made; and
- S_6: Lawsuits against the City.

It is clear that city officials have a certain amount of control over the action attributes, while state attributes cannot be directly controlled. The city government may want to create policies that decrease the occurrence of petty crimes and felonies, or increase government approval ratings if elections are nearing. Throughout the following chapters, we will use this setup as a running example to show how our approach can be used towards reaching such goals.   ∎

---

[1] http://data.nytimes.com/

[2] http://www.guardian.co.uk/data

Another kind of event KB that could be used is one containing information regarding the tactical level behaviors of organizations of interest, as well as characteristics of their sources of support and relations with the state in which they are based. Such data is available from efforts such as [5] and [13].

1.2 Related Work

There is substantial work in the AI-planning community on discovering sequences of actions that lead to a given outcome (usually specified as a goal condition, similar to what is done in this work), see [9] for an overview. However, AI planning assumes the effects of actions to be explicitly specified. Similarly, another related area is that of Reasoning about Actions [1, 10]; work in this area generally assumes that descriptions of effects of actions on fluent predicates, causal relationships between such fluents, and conditions that enable actions to be performed are available. In similar work, [3] uses a Bayesian reasoning method to model inferences about an agent's likely behavior based on external observations.

Our work approaches a problem that at first seems quite similar to AI planning in a fundamentally different and data-driven way, but making assumptions that are quite different:

1. There is a very large number of actions to choose from, given by the set of all possible ways in which the values of action attributes can be changed;
2. Actions attempting to change the value of action attributes are taken more or less in parallel, and all attempted changes succeed probabilistically depending on the *entire set* of attempted changes; and
3. The effects of the changed parameters on the state can only be determined by appeal to past data.

Research within the Machine Learning community on the problem of classification [8] is also related to our endeavor. The main differences between that research and our own is that we are not only interested in classifying situations in past data (this is actually aided by the fact that goal conditions are provided), but in how to *arrive once again at similar situations*. As we have seen, this also involves analyzing costs of performing actions and their probabilities of success. In Chap. 3, we will consider the use of classification algorithms for the underlying implementation of one of the main components of our formalism.

Another related approach is that of case-based reasoning (CBR) [7], where the knowledge base consists of a collection of so-called *cases* that were stored as specific examples of how a problem was solved in the past. In CBR, new problems are solved by iterating through the following steps: *retrieval* of cases that refer to situations that are similar to the present one; *reuse* of such cases in the context of the present situation (this step may involve adapting the solutions); and *store* the new case in the knowledge base in order to apply it again in the future. For instance, a CBR approach to home repairs in the presence of a lamp that won't turn on might have a set of repairs that have worked in the past: change the light bulb, make sure that it's plugged in, check the wiring, etc. Though our approach also relies on past experience in order to find a solution, the main difference with CBR is that event DBs do not consist of "recipes"; as we have seen, different effect estimators will produce different solutions given the same input data.

Finally, another approach that is closely related to this one is that of abductive queries in probabilistic logic-based models [11, 12, 15]. In that research line, a model is assumed to exist describing the behavior of an agent of interest (which can be an individual or a group) in the form of an *action probabilistic (ap) logic program* [2, 6, 14] containing rules of the form:

$$A : [\ell, u] \leftarrow C_1 \wedge \ldots \wedge C_n,$$

which can be read as follows:

> If the environment in which entity E operates currently satisfies conditions $C_1, \ldots, C_n$, then the probability that E will take some Boolean combination of actions A is between ℓ and u.

Similar to our approach, abductive queries consist of a probabilistic goal describing some desired combination of actions and probability bounds with which they should be entailed from the program. The model also contains functions describing the cost of changing the environment, the effect that such changes will have on the entity being modeled, and how likely the transition is to succeed. Abductive queries in *ap*-programs therefore correspond to a *model-heavy* approach to solving the problem that we are now proposing to solve in a *data-driven* manner. The advantage to the model-heavy approach is that answers to the queries can be explained to the user by means of appealing to the model itself; on the other hand, the data-driven approach proves to be much more scalable (cf. Chap. 5).

References

1. Chitta Baral and Le chi Tuan. Reasoning about actions in a probabilistic setting. In *AAAI 2002*, pages 507–512. AAAI Press, 2002.
2. Matthias Broecheler, Gerardo I. Simari, and V.S. Subrahmanian. Using histograms to better answer queries to probabilistic logic programs. In Patricia Hill and David Warren, editors, *Proceedings of ICLP 2009*, volume 5649, pages 40–54. Springer Berlin / Heidelberg, 2009.
3. C. Baker, J. Tenenbaum, and R. Saxe. Bayesian models of human action understanding. *Advances in neural information processing systems*, 18:99, 2006.
4. J.L. Davies and T.R. Gurr. *Preventive Measures: Building Risk Assessment and Crisis Early Warning Systems*. Rowman and Littlefield, 1998.
5. Center for International Development and Conflict Management. Minorities at risk organizational behavior dataset, minorities at risk project, 2008. Retrieved from http://www.cidcm.umd.edu/mar.
6. Samir Khuller, Maria Vanina Martinez, Dana Nau, Gerardo Simari, Amy Sliva, and VS Subrahmanian. Computing most probable worlds of action probabilistic logic programs: Scalable estimation for $10^{30,000}$ worlds. *Annals of Mathematics and Artificial Intelligence*, 51(2–4):295–331, 2007.
7. J. Kolodner and C.B. Reasoning. Morgan kaufmann. *San Mateo, CA*, 1993.

8. Tom M. Mitchell. *Machine Learning*. McGraw-Hill, New York, 1997.
9. Dana Nau, Malik Ghallab, and Paolo Traverso. *Automated Planning: Theory & Practice*. Morgan Kaufmann, San Francisco, CA, USA, 2004.
10. Judea Pearl. Reasoning with cause and effect. *AI Mag.*, 23(1):95–111, 2002.
11. Gerardo I. Simari, John P. Dickerson, and V.S. Subrahmanian. Cost-based query answering in probabilistic logic programs. In *Proceedings of SUM 2010*. LNCS, Springer-Verlag, 2010.
12. Gerardo I. Simari, John P. Dickerson, Amy Sliva, and V.S. Subrahmanian. Parallel abductive query answering in probabilistic logic programs. *ACM Transactions on Computational Logic, In Press*, 2013.
13. Jana Shakarian. The CMOT Codebook, available from the Laboratory for Computational Cultural Dynamics, University of Maryland Institute for Advanced Computer Studies, University of Maryland, College Park, MD 20742, USA. Extended and revised by Schuetzle, B. and Nagel, M., 2012.
14. Gerardo I. Simari, Maria Vanina Martinez, Amy Sliva, and V.S. Subrahmanian. Focused most probable world computations in probabilistic logic programs. *Annals of Mathematics and Artificial Intelligence*, 64(2–3):113–143, March 2012.
15. Gerardo I. Simari and V.S. Subrahmanian. Abductive inference in probabilistic logic programs. 7:192–201, July 2010.

Chapter 2
Optimal State Change Attempts

In this chapter, we formalize the notion of a state change attempt. The idea is that a state change attempt, when successfully applied to a given tuple, will change the action attributes with the hope that these changes will result in a change in the state. For instance, considering the school data described in the previous chapter, SCAs represent policies designed with a certain goal in mind; we provide an example in which a change in the action variables that decreases class size may lead to better proficiency scores.

Definition 2.1 (State Change Attempt (SCA)). A *simple state change attempt* is a triple (A_i, vf, vt) where $vf, vt \in Dom(A_i)$ for some $A_i \in \mathbf{A}$. A (non-simple) *state change attempt* (SCA for short) is a set $\{(A_{i_1}, vf_1, vt_1), \dots, (A_{i_k}, vf_k, vt_k)\}$ of simple state change attempts such that $i_j \neq i_k$ for all $j \neq k$.

When clear from context, we will refer to these concepts as *simple changes* and *changes*, respectively. Intuitively, a *simple* state change attempt modifies one attribute, while a state change attempt may modify more than one.

Definition 2.2 (Applicability of an SCA). Given a tuple t, an action attribute A_i, and $vf, vt \in Dom(A_i)$, a simple state change attempt (A_i, vf, vt) is *applicable* with respect to t if and only if $t(A_i) = vf$. The *result of applying a simple state change attempt* that is applicable with respect to t is t' where $t'(A_i) = vt$ and for state attribute $S_j \neq S_i$, $t'(A_j) = t(A_j)$. We use $\gamma(t, (A_i, vf, vt))$ to denote t', the tuple resulting from the application of SCA (A_i, vf, vt) to t.

A state change attempt $SCA = \{(A_{i_1}, vf_1, vt_1), \dots, (A_{i_k}, vf_k, vt_k)\}$ is *applicable* with respect to t if and only if all (A_{i_j}, vf_j, vt_j) for $1 \leq j \leq k$ are applicable with respect to t.

Example 2.1. A simple state change attempt with respect to the school data from Example 1.1 could be $a_1 = (A_1, 8,700, 8,850)$, indicating that funding is increased from \$8,700 to \$8,850 per student, or $a_2 = (A_2, 62.3, 65)$ indicating that salaries are increased from 62.3 to 65% of the budget. Let $SCA = \{a_1, a_2\}$ be a state change attempt. If we assume that the values of the action attributes in the current

A. Parker et al., *Data-driven Generation of Policies*, SpringerBriefs in Computer Science, DOI 10.1007/978-1-4939-0274-3_2, © The Author(s) 2014

environment are $t = (8,700, 64, 7, 3.2)$, then a_1 is applicable with respect to t, but a_2 is not. The result of applying a_1 to t is $\gamma(t, (A_1, 8,700, 8,850)) = t' = (8,850, 64, 7, 3.2)$.   ∎

Example 2.2. Looking now at the city government data from Example 1.2, a simple state change attempt could be $a_3 = (A_3, 6.5, 6.0)$, indicating that personal income tax is lowered from 6.5 to 6%, or $a_4 = (A_{10}, 15, 21)$, indicating that the funding for city hospitals is increased from \$12M per year to \$15M per year. Let $SCA = \{a_3, a_4\}$ be a state change attempt. Assuming that the values of the action attributes in the current environment correspond to the values of t_1 in Fig. 1.2:

$$t_1 = (2,500, \ 5.0, \ 6.5, \ 5.5, \ 350, \ 120, \ 70, \ 450, \ 430, \ 12, \ 150, \ 950)$$

then both a_3 and a_4 are applicable. The result of applying SCA is $\gamma(t, SCA) =$

$$(2,500, \ 5.0, \ 6.0, \ 5.5, \ 350, \ 120, \ 70, \ 450, \ 430, \ 15, \ 150, \ 950).$$ ∎

The result of applying a state change attempt is therefore the result of applying each simple change. However, these changes do not occur without cost.

Definition 2.3 (Cost of a simple change attempt). Let $a = (A_i, vf, vt)$ be a simple state change attempt. The *cost of attempting* a is given by a real-valued function $cost : \{A_1, \ldots, A_m\} \times \mathbb{R} \times \mathbb{R} \to \mathbb{R}$, where $cost(A_i, vf, vt)$ is the cost of changing action attribute A_i from vf to vt.

Cost functions will be highly dependent on the application domain, and we assume them to be provided by a user. The *cost* of an attempt, $cost(SCA) = \sum_{a \in SCA} cost(a)$, is the sum of the costs of the simple state change attempts in SCA.

Example 2.3. Consider the same simple changes $a_1 = (A_1, 8,700, 8,850)$ and $a_2 = (A_2, 62.3, 65)$ from Example 2.1, and a third simple change $a_3 = (A_4, 3.8, 3.9)$ (i.e., increment the proficiency increase target from 3.8 to 3.9). A possible cost function could be defined in terms of monetary cost, in which: $cost(a_1) = 150 * s$ (where s is a constant set to the number of students affected), $cost(a_2) = 2.7 * A_1$, and $cost(a_3) = 0$ (no monetary cost associated with changing the proficiency increase target).   ∎

Example 2.4. A different cost function may be defined for the city government dataset, perhaps incorporating the political capital or risk necessary to undertake government actions. Consider again the state change attempt $SCA = \{a_3, a_4\}$ from Example 2.2 where $a_3 = (A_3, 6.5, 6.0)$ and $a_4 = (A_{10}, 15, 21)$. Lowering income tax may require striking deals with opposing political parties, and the effect on the budget may be negative at first, thereby making a_3 both financially and politically costly: $cost(a_3) = 200 + 0.5 * d$, where d indicates how many half percentage points the tax is lowered (1 for a_3). Influencing the city's policy towards a higher hospital budget may require political actions that are less costly given the expected benefits of such a change: $cost(a_4) = 5 + 2 * d'$, where d' indicates the number of \$100K increments are being proposed (60 for a_4).   ∎

To this point, we have looked at state change attempts that are always successful. However, in general, we cannot expect this to be the case—the funding per student may not change simply because one attempted to change it. We will assume state change attempts are only probabilistically successful—they only induce the change attempted according to a specified probability. Further, we will assume that the probability of any simple change occurring successfully depends on the entire set of changes attempted. For instance, when attempting to increase the proficiency target alone, one may expect a relatively small probability of the change actually occurring—the teacher's union is unlikely to accept an increase in their expected performance with no additional compensation. However, when attempting to increase the proficiency target along with an increase in teacher salaries, as in Example 2.3, there will be a higher probability that both changes will actually occur.

In general, such a change can cause four possible outcomes. It may be that both the proficiency target increases and the salaries increase, or that the proficiency target increases while the salaries do not change, or that the proficiency target stays the same while the salaries increase, or that both do not change. The probability of these outcomes is dependent on the total state change attempt: the school board is more likely to accept a salary increase along with an increase in the proficiency target while the teacher's union is more likely to accept a proficiency target increase as long as it also comes with a salary increase.

Example 2.5. Consider the situation described in Example 2.3. Here the state change attempt a_2 increases teacher salaries from 62.3 to 65%. On its own, this policy may anger taxpayers (who would foot the bill for the increase) and may only have a 10% probability of succeeding. Likewise, increasing per student funding might have a 15% probability of success. However, if the taxpayers happen to be willing to increase teacher salaries, then they will also tend to approve per student funding increases, perhaps leading to a joint probability of 9% that both of these will occur when attempted together.

Similar probability dependencies may be seen in the case of state change attempts in Example 2.4. The city government may be more likely to increase funding for the police department if other public safety increases are already being proposed (such as an increment of the budget for street lighting). Similarly, any policy involving an increment in funding is more likely to succeed along with an increment in the overall budget, as opposed to in combination with a reduction in funding for a different item.   ∎

Let *SCA* and $SCA' \subseteq SCA$ be state change attempts; suppose we have the conditional probabilities $pOccur(SCA'|SCA)$—this is the probability that only the actions in SCA' occur given that SCA is attempted. Such probabilities can either be derived from historical data or be explicitly stated by a user. When we say that a state change attempt *SCA* is "attempted" for a tuple t describing the current situation, this means that each $SCA' \subseteq SCA$ has the chance $pOccur(SCA'|SCA)$ of being successful, i.e., of having $\gamma(t, SCA')$ be the resulting tuple.

Example 2.6. Consider once again some state change attempts from Example 2.3: $a_1 = (A_1, 8{,}700, 8{,}850)$, and $a_2 = (A_2, 62.3, 65)$. If we apply the state change attempt $\{a_1, a_2\}$, there are four possible outcomes:

1. a_1 and a_2 could occur, increasing the funding per student to \$8,850 as well as increasing teachers' salaries to 65% of the budget;
2. a_1 could occur without a_2, increasing the funding per student without changing the budget teacher's salaries fraction of the budget;
3. a_1 could fail while a_2 occurs, leaving funding per student at \$8,700 but increasing teacher salaries; or
4. Both fail, leaving things as they are.

The probabilities of these outcomes are respectively given by the state change conjunction strategy as follows:

$pOccur(\{a_1, a_2\} \mid \{a_1, a_2\}), pOccur(\{a_1\} \mid \{a_1, a_2\}),$

$pOccur(\{a_2\} \mid \{a_1, a_2\}),$ and $pOccur(\emptyset \mid \{a_1, a_2\}).$ ∎

In the next section we will present effect estimators, which constitute an important part of the policy generation process.

2.1   Effect Estimators

The goal in this work is to allow an end user to take an event KB $\mathcal{K}$ and a goal G (some desired outcome condition on state attributes) that the user wants to achieve and find an SCA that "optimally" achieves goal G in accordance with some objective function (such as maximizing the probability of goal G being achieved while minimizing cost). We assume without loss of generality that all goals are expressed as standard conjunctive selection conditions [3] on state attributes. We now define *effect estimators*.

Definition 2.4 (Effect Estimator). For action tuple t and goal G, an *effect estimator* is a function $\varepsilon(t, G) \to [0, 1]$ that maps a tuple and a goal to a probability $p \in [0, 1]$.

Intuitively, $\varepsilon(t, G)$ specifies the conditional probability of goal G holding given that we are in a situation where the action attributes are as specified in t. This quantity can be estimated in many ways, some of which will be investigated later on in Chap. 3. As an initial example, one can imagine using some machine learning algorithm as an effect estimator to determine how often a school's reading score is above some number k given that the student-teacher ratio, funding per student, teacher salaries, etc. have some specific values.

2.2 State Change Effectiveness

We assume an environment where, just because a state change attempt is performed, it is not necessarily the case that all parts of the state change attempt will actually accomplish the attempted change. When one attempts to change the situation via a state change attempt *SCA*, any subset of *SCA* may succeed. For instance, if a certain policy involves decreasing the student/staff ratio and increasing the funding per student, it may be that the student/staff ratio increases as expected, but that the funding per student remains the same. Thus, to truly gauge the effectiveness of a state change attempt, we must consider the probability of each subset of the attempt occurring.[1]

Definition 2.5 (State Change Effectiveness). The probability of a state change attempt $SCA = \{(A_{i_1}, vf_1, vt_1), \ldots, (A_{i_k}, vf_k, vt_k)\}$ satisfying goal G when applied to the action tuple t is

$$pEff(t, G, SCA, \varepsilon) = \sum_{SCA' \in \mathscr{P}(SCA)} pOccur(SCA' \mid SCA)\varepsilon(\gamma(t, SCA'), G).$$

The computation of $pEff(t, G, SCA, \varepsilon)$ works by summing over all the state changes that may occur given that *SCA* is attempted: since any subset of *SCA* can occur, this summation ranges over $SCA' \subseteq SCA$. For each SCA' that may occur, one multiplies its probability of occurring given that *SCA* was attempted ($pOccur(SCA' \mid SCA)$) times the effectiveness of the given attempt according to ε (recall that $\gamma(t, SCA')$ is the action tuple resulting from the application of the state change SCA' to the original action tuple t). The following result shows that for arbitrary effect estimators, computing state change effectiveness is intractable.

Proposition 2.1. *For action attributes* **A**, *state attributes* **S**, *condition G, state change attempt SCA, action tuple t, and effect estimator ε, deciding whether $pEff(t, G, SCA, \varepsilon) > 0$ is NP-hard. Furthermore, if $\varepsilon(.)$ can be computed in polynomial time with respect to the action schema, the problem is NP-complete.*

Proof. Membership In NP: We show that deciding if $pEff(t, G, SCA) > 0$ is in NP with a witness $SCA' \subseteq SCA$ such that $pOccur(SCA') \cdot (1 - pOccur(SCA \setminus SCA')) > 0$. Since $pEff(s, G, SCA, \varepsilon)$ is a sum of non-negative terms, such an SCA' must exist when $pEff(t, G, SCA, \varepsilon) > 0$, and this can be checked in polynomial time with respect to $|\mathscr{A}|$ if $\varepsilon(.)$ can be computed in this time.

NP-hardness: We show by reduction from the NP-complete subset-sum problem, whereby we are given a finite set of integers I and an integer c and are asked to decide if there is a subset $I' \subseteq I$ such that $\sum_{i \in I'} i = c$ [1]. Let $I = \{i_1, \ldots, i_n\}$ and let $|\mathbf{A}| = \{A_1, \ldots, A_n\}$ with $Dom(A_i) = \{0, i_i\}$ for $A_i \in \mathbf{A}$. Let action

---

[1] In this work, we assume that each simple change attempt either succeeds or fails completely, i.e., no partial effects can occur.

tuple t be the all-zero n-tuple. Define $SCA = \{(A_j, 0, i_j) \mid 1 \le j \le n\}$, and define all $pOccur(SCA')$ to be zero unless $\sum_{A \in \mathbf{A}} \gamma(t, SCA')(A) = c$, in which case $pOccur(SCA') = 1/2$. let G be the condition $\top$, ε is a constant function that returns 1 for any KB and conditions, and suppose $\mathcal{K}$ is the empty event KB. Under these conditions $pEff(t, G, SCA, \varepsilon) > 0$ if and only if there is a subset of I summing to c.

($\Rightarrow$): Suppose $pEff(t, G, SCA, \varepsilon) > 0$ to show there is a subset of I summing to c. Since $pEff(t, G, SCA, \varepsilon) > 0$, there is $SCA' \subseteq SCA$ such that:

$$pOccur(SCA')(1 - pOccur(SCA \setminus SCA')) > 0.$$

This implies that $pOccur(SCA') > 0$, which in turn implies that for $t' = \gamma(t, SCA')$, $\sum_{A \in \mathbf{A}} t'(A) = c$. Since t is the zero n-tuple, all non-zero $t(A)$ result from changes in SCA', so we have that $\sum_{(A_j, 0, i_j)} i_j = c$. This gives the set $I' = \{i_j \mid (A_j, 0, i_j) \in SCA'\}$ which describes a subset of I whose sum is c.

($\Leftarrow$): Let $I' \subseteq I$ be the subset of I such that $\sum_{i_j \in I'} i_j = c$. Now consider state change $SCA' = \{(A_j, 0, i_j) \mid i_j \in I'\}$. Clearly $SCA' \subseteq SCA$ and $(\gamma(t, SCA'), \cdot) \in \sigma_G(\mathcal{K})$, so

$$pOccur(SCA') \cdot (1 - pOccur(SCA \setminus SCA')) \qquad (2.1)$$

will be a term in the sum defining $pEff(t, G, SCA, \varepsilon)$. Since $\sum_{i_j \in I'} i_j = c$ we know that for $t' = \gamma(t, SCA')$, $\sum_{A \in \mathbf{A}} t'(A) = c$ and therefore that $pOccur(SCA') > 0$. Since $(1 - pOccur(SCA \setminus SCA'))$ is at least 0.5, this is the only term in Eq. 2.1 that may potentially be zero, proving that Eq. 2.1 is non-zero. Further, since all terms in the sum defining $pEff(t, G, SCA, \varepsilon)$ are zero or positive, this suffices to prove that $pEff(t, G, SCA, \varepsilon) > 0$. $\square$

2.3 Optimal State Change Attempts

We now present various problems related to finding optimal state change attempts or policies. We assume that we are given $\mathbf{A} = \langle A_1, \ldots, A_n \rangle$ and $\mathbf{S} = \langle S_1, \ldots, S_m \rangle$, an event KB $\mathcal{K}$, an action tuple t describing the current values of the action attributes, a goal G over $\mathbf{S}$, and functions $cost$ and $pOccur$ as mentioned earlier:

1. **The Lowest Cost SCA Problem.** Given real number κ, does there exist an applicable change attempt SCA such that $cost(SCA) \le \kappa$ and $pEff(t, G, SCA, \varepsilon) > 0$?
2. **The Highest Probability SCA Problem.** Given a real number $p \in [0, 1]$, does there exist a change attempt SCA such that $pEff(t, G, SCA, \varepsilon) \ge p$?
3. **The Optimal Threshold Effectiveness Problem.** Given a threshold $p \in [0, 1]$, and cost k, does there exist a change attempt SCA such that $pEff(t, G, SCA, \varepsilon) \ge p$ and $cost(SCA) \le k$? This problem is the result of combining both the Highest Probability and Lowest Cost problems stated above.

4. **The Limited Cardinality SCA Problems.** Given a positive integer h, does there exist an *SCA* such that $|SCA| \leq h$ and *SCA* satisfies one of the conditions from the problems above? For instance, the limited cardinality highest probability SCA will, given a real number $p \in [0, 1]$, tell if there exists a change attempt *SCA* such that $pEff(t, G, SCA, \varepsilon) \geq p$ and $|SCA| \leq h$.

All of these problems are stated as decision problems that ask whether an *SCA* satisfying certain conditions exists. Search problems, to *find* such an SCA, can be analogously stated. We refer to any state change attempt that is a solution to one of these problems (say, problem P) as an *optimal state change attempt* (OSCA, for short) with respect to P.

Theorem 2.1. *If the effect estimator used can be computed in PTIME, the decision problems associated with the different definitions of optimal state change attempts belong to the following complexity classes:*

1. *The Lowest Cost SCA problem is NP-complete.*
2. *The Highest Prob. SCA problem is #P-hard and in PSPACE.*
3. *The Optimal Threshold Effectiveness Problem is #P-hard and in PSPACE.*
4. *All Limited Cardinality SCA problems are in PTIME.*

Proof. We prove each part in turn:

1. *Membership in NP*: Let ε be an effect estimator that can be computed in polynomial time with respect to $|\mathbf{A}|$. A witness change attempt *SCA*, along with $SCA' \subseteq SCA$ such that $cost(SCA) \leq \kappa$, $pOccur(SCA')(1 - pOccur(SCA \setminus SCA')) \cdot \varepsilon(t' = \gamma(t, SCA'), G) > 0$, and $(\gamma(t, SCA'), \cdot) \in \sigma_G(\mathcal{K})$ (implying $pEff(t, G, SCA, \varepsilon) > 0$) can be verified in polynomial time with respect to $|\mathbf{A}|$.

   NP-Hardness: The NP-hardness proof from Proposition 2.1 can be extended for this purpose by simply assuming that the only possible values in $Dom(A_i)$ are 0 and i_j (cf. the reduction in the proof) and assigning κ to be one greater than the sum of state change attempt costs. Therefore, any subset of *SCA* as defined for which there exists a subset of I summing to c can be seen as a state change attempt with the required property.

2. *# P-hard*: Let F be a SAT formula with variables $v_0, \ldots, v_n$, and N be a number. The problem of determining if the number of solutions to F is greater than or equal to N is #P-complete [2, 4]. We let there be an action attribute A_i for $i = 0, 1, \ldots, n$, with domain $Dom(A_i) = \{0, 1\}$. Let $t(A_i) = 0$ in the action tuple t for all A_i. All applicable simple state change attempts have the form $(A_i, 0, 1)$. Define the cost function to always return 0. We let *pOccur* always be 0.5, and let $\varepsilon(C_1, C_2)$ be one if $C_2 = F$ and C_1 exactly specifies a tuple t', where t' satisfies F, and zero otherwise. Define p to be $N \cdot 0.25$. The number of solutions to F is greater than or equal to N if and only if there is an applicable state change attempt *SCA* such that the cost of *SCA* is less than or equal to 0 and $pEff(t, F, SCA, \varepsilon) \geq p$.

Membership in PSPACE: All possible state change attempts can be checked by keeping track of only one at a time, which can clearly be done within the polynomial space constraint.

3. Analogous to part 2.
4. Since the size of the state change attempt is at most h, the space of possible state change attempts is bounded by a polynomial in $|\mathbf{A}|$. To see why this is the case, it suffices to recall that each $A_i \in \mathbf{A}$ has a *finite* domain $dom(A_i)$, that a state change attempt can specify at most one simple state change attempt per attribute in $\mathbf{A}$, and that $\mathscr{C}_h^n \in O(n^h)$, where $n = |\mathbf{A}|$. Therefore, in the worst case we have to check $\sum_{A \subseteq \mathbf{A}, |A| \leq h} \left(\prod_{A_i \in A} dom(A_i) \right)$ SCAs, which is a quantity in $O(n^h)$. $\square$

In summary, in the proof of Theorem 2.1, all NP-hard reductions use the results of Proposition 2.1, the #P-hard reductions use #*SAT* (the language $\{\langle F, n \rangle\}$, where F is a formula with exactly n solutions), membership in NP is shown by providing a witness, and membership in PSPACE and PTIME are shown by sketching algorithms, with the necessary properties, which are discussed in more detail in Sect. 2.4. In all cases, the source of the complexity can be seen to stem from $|\mathbf{A}|$.

2.4   Basic Algorithms for Computing OSCAs

We begin by giving an explicit polynomial time algorithm for solving the limited cardinality SCA problems; this corresponds to the procedure discussed in the proof of Theorem 2.1. We will then show how to extend this algorithm to solve all the problems posed in the last section.

This algorithm works by first enumerating each possible state change attempt with size at most h, then choosing the one which solves the appropriate problem. As discussed in the proof of Theorem 2.1, since there are only $O(|\mathbf{A}|^h)$ such state change attempts, this algorithm runs in PTIME with respect to the number of action attributes $|\mathbf{A}|$. The algorithm for enumerating state change attempts of size at most h along with their cost and probability of effectiveness is given as Algorithm 1, and we can show that this algorithm runs in time in $O(|\mathbf{A}|^h)$.

Proposition 2.2. *Algorithm 1 runs in time in $O(|\mathbf{A}|^h)$ and returns all (SCA, c, ef) where $|SCA| \leq h$, $c = cost(SCA)$ and $ef = pEff(t, G, SCA, \varepsilon)$.*

Proof. The size of R is at most $(|\mathbf{A}| \cdot \max_i(|dom(A_i)|))^h$, and Algorithm 1 terminates in $O(|R|)$ steps. Since we consider $\max_i(|dom(A_i)|)$ to be a constant, this is $O(|\mathbf{A}|^h)$. Further, to see that R is correct, clearly c and ef are correct for each (SCA, c, ef) (see line 10), so it remains to show that all SCA of size $\leq h$ are included in R. Consider any state chance attempt SCA to show there is a (SCA, c, ef) in R. We show this by induction on $|SCA|$. As a base case, when

---

Algorithm 1 limitedSCASet(t, G, ε, h)

Returns the set (SCA, c, ef), where c is the cost of state change attempt SCA and ef is the probability of effectiveness of SCA

1: Let $R = \emptyset$ // the set to be returned.
2:   Add $(\emptyset, 0, pEff(t, G, \emptyset, \varepsilon))$ to R. // Initialize R with empty state change attempt.
3: **for** each $A_i \in \mathbf{A}$ **do**
4:     **for** each value $v \in dom(A_i)$ **do**
5:         **continue** if $v = t(A_i)$ // Go to next value, t won't be changed by this SCA.
6:         // iterate over all members of R, growing those that are small enough.
7:         **for** each $(SCA, c, ef) \in R$ **do**
8:             **continue** if $|SCA| = h$.
9:             Let $SCA' = SCA \cup \{(A_i, t(A_i), v)\}$.
10:            Let c' be the cost of SCA' and ef' be $pEff(t, G, SCA', \varepsilon)$.
11:            Add (SCA', c', ef') to R.
12:        **end for**
13:    **end for**
14: **end for**
15: **return** R

$|SCA|$ is zero, (SCA, c, ef) is in R. Supposing all SCA of size k are in R to show that any SCA_{k+1} of size $k + 1$ is in R. Let $SCA_k \cup \{(A_i^*, vf, vt)\}$ be SCA_{k+1}. Because $|SCA_k|$ has size k, there is (SCA_k, c, ef_k) in R. Therefore when we run line 11 with $SCA' = SCA_k$, $A_i = A_i^*$ and $v = vt$, then $(SCA_{k+1}, c_{k+1}, ef_{k+1})$ will be added to R. Thus all SCA of size less than or equal to $k + 1$ will be in R as (SCA, c, ef) with correct c and ef.   □

Using Algorithm 1, we can now compute solutions to each of the limited cardinality SCA problems.

- **Lowest Cost Limited SCA Algorithm**: For cost threshold κ, let R be limitedSCASet(t, G, ε, h), eliminate all $(SCA, c, 0)$ from R, then eliminate all (SCA, c, ef) from R where $c > \kappa$. Return true if R is non-empty, false otherwise. $(SCA, c, ef) \in R$ has minimal c.
- **Highest Probability Limited SCA Algorithm**: For probability threshold p, let R be limitedSCASet(t, G, ε, h), eliminate all (SCA, c, ef) where $ef < p$ and return true if R is non-empty, false otherwise.
- **Optimal Threshold Effectiveness Limited SCA Algorithm**: For probability threshold p and cost threshold κ, let R be limitedSCASet(t, G, ε, h), and eliminate all (SCA, c, ef) from R where either $c > \kappa$ or $ef < p$. Return true if R is non-empty, false otherwise.

Each of those algorithms correctly computes the associated decision problem, as a corollary of Proposition 2.2.

Corollary 2.1. *Each of the following algorithms correctly computes the associated decision problem: Lowest Cost Limited SCA Algorithm, Highest Probability Limited SCA Algorithm, and Optimal Threshold Effectiveness Limited SCA Algorithm.*

To extend this technique to the non-limited, general versions of the various problems, one simply needs to solve the limited version of the problem with h equal to $|\mathbf{A}|$; however, in this case the algorithm will no longer run in polynomial time (cf. Theorem 2.1).

Again, each of the corresponding algorithms will correctly compute the associated decision problem, as a corollary of Proposition 2.2.

References

1. Thomas H. Cormen, Charles E. Leiserson, Ronald L. Rivest, and Clifford Stein. *Introduction to Algorithms, Second Edition*. The MIT Press, September 2001.
2. Oded Goldreich. *Computational Complexity: A Conceptual Perspective*. Cambridge University Press, 1 edition, April 2008.
3. Jeffrey D. Ullman. *Principles of Database and Knowledge-Base Systems, Volume I*. Computer Science Press, 1988.
4. L.G. Valiant. The complexity of computing the permanent. *Theoretical Computer Science*, 8(2):189–201, 1979.

Chapter 3
Different Kinds of Effect Estimators

In this chapter we introduce several sorts of effect estimator, which yield the likelihood of a given action tuple satisfying a given goal condition G. An effect estimator essentially answers the question: "if I succeed in changing the environment in this way, what is the probability that the environment satisfies my goal?". We also present the TOSCA algorithm, an optimized approach to computing optimal state change attempts when using a special kind of effect estimator.

3.1 Learning Algorithms as Effect Estimators

In this section, we describe how to take any supervised learning algorithm [2] (i.e., neural nets, decision trees, case based learning, etc.) and apply it to the event knowledge base $\mathscr{K}$ to get an effect estimator. Supervised learning algorithms require training data with categorization for that data as positive or negative instances of a given category. From that training data, they construct a *classifier*, or a procedure that classifies future cases—even those not already seen.

From a given goal condition G and knowledge base $\mathscr{K}$, we can construct training data to which we can apply any standard machine learning technique. To do this, we categorize each member of $\mathscr{K}$ according to G: that is, if it satisfies G then the tuple is a positive instance, while if it does not the tuple is a negative instance. We then use only the action portion of the tuple along with these categorizations to train a decision tree, a neural network, a support vector machine or some other classifier. The resulting classifier will be the effect estimator—it will tell for any given action tuple if the resulting state attributes are likely to be a positive instance (satisfying the goal G) or not.

We abstractly model a machine learning algorithm as a *learner*, which, given the appropriate information, will produce a *classifier*.

A. Parker et al., *Data-driven Generation of Policies*, SpringerBriefs in Computer Science, DOI 10.1007/978-1-4939-0274-3_3, © The Author(s) 2014

Definition 3.1 (Classification Algorithm). For event KB $\mathcal{K}$ and goal condition G, a classification algorithm is a function learner : $(\mathcal{K}, G) \mapsto$ classifier, where classifer is a function from action tuples to the interval $[0, 1]$.

Example 3.1. A neural network fits this definition in the following way [3]: we first define learner to be a function that generates a neural network with input nodes for each action attribute and exactly one output node with a domain of $[0, 1]$. The learner function then trains the network via backpropogation according to $\mathcal{K}$ and G, where those tuples in $\mathcal{K}$ that satisfy G are positive instances (expecting the output node to have value 1) and those tuples in $\mathcal{K}$ that do not satisfy G are negative instances (expecting the output node to have a value of 0). The resulting network is the classifier function, and will, given a set of values for the action attributes, return a value in the interval $[0, 1]$. ■

We can use a classification algorithm to create a *learned effect estimator*.

Definition 3.2. Given a classification algorithm *learner*, a *learned effect estimator* is defined to be $\varepsilon_{lrn}(learner, \mathcal{K})(t, G)$. We define the *learned effect estimator* to return $learner(\mathcal{K}, G)(t)$.

Example 3.2. If we consider the learning algorithm to be C4.5, then $\varepsilon_{lrn}(C4.5, \mathcal{K})$ (t, G) will first construct a decision tree T according to samples from $\mathcal{K}$ classified according to G. Then we will query the decision tree T to classify the action tuple t, and this classification will be our return value—1 if t is classified the same as tuples satisfying G, and 0 if not (recall, t is an action tuple and the goal G is a formula over state attributes, so there is no way to check if t satisfies G directly). ■

3.2 Data Selection Effect Estimators

In this section we examine the special case of an effect estimator that uses selection operations in a database to create an estimation. For our purposes, selection operations will be denoted $\sigma_G(\mathcal{K})$, where $\mathcal{K}$ is an event KB and G is some goal condition on the state tuples. The $\sigma_G(\mathcal{K})$ operation returns the subset of $\mathcal{K}$ satisfying the condition G.

Definition 3.3 (Data Selection Effect Estimator). For goal G and action tuple t, a data selection effect estimator is a function that takes an event knowledge base $\mathcal{K}$ as input and returns an effect estimator: $\varepsilon^* : \mathcal{K} \mapsto (t, G) \mapsto p$, where $p \in [0, 1]$. We require the following conditions to hold:

1. It is possible to implement ε^* with a fixed number of selection operations on $\mathcal{K}$, and
2. $\varepsilon^*(\mathcal{K})(t, G) = 0$ whenever there does not exist any tuple in $\mathcal{K}$ whose action attributes match t.

A data selection effect estimator differs from a normal effect estimator in that it depends explicitly on selection from event KB $\mathcal{K}$. While data selection effect estimators are limited to using only selection operators, we will see that there are many ways to specify the relationship between G and the situation described by t under this limitation (cf., for instance, Definitions 3.4, and 3.5).

In the following, we will slightly abuse the notation used for selection operators in databases by writing $\sigma_t(\mathcal{K})$ to denote the selection of all the tuples in $\mathcal{K}$ that have the values described by t for the corresponding attributes.

Definition 3.4 (Data Ratio Effect Estimator). The data ratio effect estimator returns the fraction of the time that G holds out of all times when the attributes in t are matched.

$$\varepsilon_r^*(\mathcal{K})(t, G) \stackrel{def}{=} \begin{cases} \frac{|\sigma_{t \wedge G}(\mathcal{K})|}{|\sigma_t(\mathcal{K})|} & : |\sigma_t(\mathcal{K})| > 0 \\ 0 & : |\sigma_t(\mathcal{K})| = 0 \end{cases}$$

The data ratio effect estimator returns the marginal probability of G occurring given that the values specified by the action tuple t occur.

Example 3.3. Suppose we have a school metrics database containing only three columns: class size, teacher salary and graduation rate. The class size and teacher salary are action attributes, while the graduation rate is a state attribute. We want to determine from the data what fraction of the time a graduation rate is at least 95% for an average class size of 20 and an average teacher salary of $60,000. According to ε_r^*, this fraction is the fraction of tuples in the database with class size 20 and teacher salary $60,000 that have a graduation rate over 95% divided by the total number of tuples in the database with class size 20 and teacher salary $60,000.   ∎

Example 3.4. We can also look at how a data ratio effect estimator would operate on the city government database. Suppose we only have the columns *Funding for Police Department*, *Funding for Street Lighting*, and *Petty Crimes*, where the two former are action attributes and the latter is a stat attribute. In this case, we may want to determine what fraction of the time the incidence of petty crimes is above 125 occurrences when funding for the police department is below $400M and the funding for street lighting is below $85K. Using ε_r^*, we divide the number of tuples in the database where $A_5 \leq 400$, $A_7 \leq 85$, and $S_3 \geq 125$ by the number of tuples where $A_5 \leq 400$ and $A_7 \leq 85$. For the event database in Fig. 1.2, this yields $1/3$.   ∎

One important feature of the data ratio effect estimator is that when there is no information in the database on a given tuple, the data ratio effect estimator assumes the tuple to be a negative instance. This will allow the quick elimination of possibilities not contained in the database, and will reduce the search space needed to compute optimal state change attempts.

Further examples of data selection effect estimators include cautious or optimistic ratio effect estimators, which take the confidence interval into account.

Definition 3.5 (Cautious Ratio Effect Estimators). The cautious ratio effect estimator returns the probability of G given t to be the low end of the 95% confidence interval:

$$\varepsilon^*_{c95}\mathcal{K}(t, G) \stackrel{def}{=} \varepsilon^*_r(\mathcal{K})(t, G) - 1.96 \cdot \sqrt{\frac{\varepsilon^*_r(\mathcal{K})(t, G)(1 - \varepsilon^*_r(\mathcal{K})(t, G))}{|\sigma_t(\mathcal{K})|}}$$

(if $\sigma_t(\mathcal{K})$ is empty, $\varepsilon^*_{c95}(\mathcal{K})(t, G)$ is defined to be zero).

There is a whole class of cautious ratio effect estimators: one for every confidence level (90, 80, 99%, etc.). There are also optimistic ratio effect estimators, which return the high end rather than the low end of the confidence interval.

Since data selection effect estimators are computed via a finite number of selection operations, effect estimators can always be computed in time in $O(|\mathcal{K}|)$. The complexity of finding SCAs changes when we insist on using data selection effect estimators. Problems that were NP-complete or #P-hard with respect to the size of the action schema are polynomial in $|\mathcal{K}|$ when only data selection effect estimators are allowed, as discussed in the next section.

3.3 Computing OSCAs with Data Selection Effect Estimators

Using data selection effect estimators, we can devise specific algorithms for finding optimal state change attempts. In this section we use only the data ratio effect estimator (Definition 3.4). Algorithm 2 presents the **DSEE_OSCA** algorithm to solve the optimal threshold effectiveness problem.

The **DSEE_OSCA** algorithm works by selecting all tuples in the event KB $\mathcal{K}$ satisfying the goal condition, then adding the pair (SCA, f) to a data structure $Dat1$ where f is the chance that SCA, when successful, results in a state satisfying the goal G (i.e., $\varepsilon^*_r(\mathcal{K})(t, G)$). In the next loop, two things happen: (i) f is multiplied by the probability that SCA is successful, and (ii) we iterate through all state change attempts and sum the probability of occurrence of each subset of SCA with that subset's probability of satisfying the goal G, adding the result to data structure $Dat2$. At this point $Dat2$ contains pairs (SCA, ef), where ef is the probability of effectiveness of SCA according to Definition 2.5. The algorithm then prunes all state change attempts without sufficiently high probabilities of effectiveness, and returns the one with the lowest cost.

The following propositions state that the **DSEE_OSCA** algorithm is correct, as well as analyze its running time.

Proposition 3.1. *Algorithm 2 computes a state change attempt SCA with the following properties:*

Algorithm 2 DSEE_OSCA(KB $\mathcal{K}$, Goal G, Action tuple env, p)

A brute force algorithm for computing a state change attempt with minimal cost and probability of effectiveness at least p

---

1: Let $Dat1 = \emptyset$ // $Dat1$ will contain state change attempts and their probability of occurrence.
2: // Iterate through all tuples satisfying G in $\mathcal{K}$.
3: **for** $t \in \sigma_G(\mathcal{K})$ **do**
4:     // Create SCA such that $\gamma(env, SCA)$ equals t on action attributes.
5:     $SCA = \{(A, env(A), t(A)) \mid env(A) \neq t(A)\}$
6:     If $(SCA, \cdot) \in Dat1$ then continue. // Already visited
7:     Let $f = \varepsilon_r^*(\mathcal{K})(t, G)$.
8:     Add (SCA, f) to $Dat1$.
9: **end for**
10: Let $Dat2 = \emptyset$
11: **for** $(SCA, f) \in Dat1$ **do**
12:     Let $nextF = pOccur(SCA|SCA) \cdot f$.
13:     **for** $(SCA', f') \in Dat1$ **do**
14:         **if** $SCA' \subsetneq SCA$ **then**
15:             $nextF = nextF + pOccur(SCA'|SCA) \cdot f'$
16:         **end if**
17:     **end for**
18:     Add (SCA, ef) where $(SCA, nextF)$ to $Dat2$.
19: **end for**
20: Remove any (SCA, ef) from $Dat2$ where $ef < p$.
21: **return** $argmin_{(SCA, ef) \in Dat2}(cost(SCA))$.

---

1. $pEff(env, G, SCA, \varepsilon_r^*(\mathcal{K})) \geq p$, and
2. There is no other applicable state change attempt SCA' such that $cost(SCA') \leq k$ and $pEff(env, G, SCA', \varepsilon_r^*(\mathcal{K})) \geq p$.

Proof. To show this, it suffices to show that in line 19 for all $(SCA, ef) \in Dat2$, $ef = pEff(env, G, SCA, \varepsilon_r^*(\mathcal{K}))$. Consider any $(SCA, ef) \in Dat2$ on that line, and note that due to the loop starting in line 9,

$$ef = \sum_{(SCA', f') \in Dat1, SCA' \subseteq SCA} f' \cdot pOccur(SCA' \mid SCA).$$

Since for all $(SCA', f') \in Dat1$, $f' = \varepsilon_r^*(\mathcal{K})(\gamma(SCA', env), G)$, this suffices to show that $ef = pEff(SCA, G, env, \varepsilon_r^*(\mathcal{K}))$.

Finally, lines 20 and 21 guarantee that the only SCAs returned are those that have probability of effectiveness at least p and cost at most k.   □

Proposition 3.2. *Algorithm 2 runs in time* $O(|\mathcal{K}|^2)$.

Proof. The running time of the algorithm can be divided into two parts. First, the loop in line 3 runs at most $|\mathcal{K}|$ times. In each iteration, we compute line 7, which takes at most $2 \cdot |\mathcal{K}|$ computations for ε_r^* (in the worst case, both select operations return the entire database). This gives a total running time of $O(|\mathcal{K}|^2)$ for the first loop. Then we get to the loop in line 11, which since $|Dat2|$ can be no larger than $|\mathcal{K}|$, will run $O(|\mathcal{K}|)$ times. The contained loop starting in line 13 will for the same reason run $O(|\mathcal{K}|)$ times, giving a run time of $O(|\mathcal{K}|^2)$ for that loop. This puts the total run time at $O(|\mathcal{K}|^2)$. $\square$

As a corollary to Propositions 3.1 and 3.2 we can arrive at the following result.

Corollary 3.1. *If the effect estimator is a data selection effect estimator, then the Lowest Cost, Highest Probability, Limited Cardinality, and Optimal Threshold Effectiveness problems can all be solved in $O(|\mathcal{K}|^2)$ time and are therefore in PTIME with respect to the number of tuples in the event KB.*

Proof. Direct consequence of Propositions 3.1 and 3.2. $\square$

Finally, we conclude this section with a more general result regarding the complexity of deciding the probability of effectiveness of a given state change attempt when the effect estimator can be computed in polynomial time.

Theorem 3.1. *For goal G, state change attempt SCA, action tuple t, and event KB $\mathcal{K}$, if the effect estimator ε^* is a data selection effect estimator then deciding if $pEff(t, G, SCA, \varepsilon^*(\mathcal{K})) > 0$ takes $O(|\mathcal{K}|^2)$ time.*

Proof. Consider Algorithm 2 and the analysis of its running time in the proof of Proposition 3.2. If ε^* is not a data selection effect estimator but can be computed in polynomial time, then the loop in line 3 can be computed in $O(|| \cdot \mathcal{E})$, where E is the cost of computing ε^*. Therefore, the total running time in this case is $O(|\mathcal{K}| \cdot E + |\mathcal{K}|^2)$ $\square$

3.4   Trie-Enhanced Optimal State Change Attempts (TOSCA)

In this section, we present the **TOSCA** algorithm that uses trie data structures [1] to improve performance of finding an optimal state change attempt. In **TOSCA**, a trie structure is used to index the event KB with the objective of reducing the search space necessary for the data selection effect estimator in the **DSEE_OSCA** algorithm; the algorithm is presented in Algorithm 2.

Preliminaries on Tries. A trie is a tree-based data structure composed of *internal nodes* and *leaf nodes*. In the following, we will present the basics of the particular adaptation of the trie data structure to our setting. An example of such a structure can be found in Fig. 3.1, which is a trie indexing the database in Fig. 3.2.

An *internal* trie node is a pair $(Atr, Edges)$, where $Atr \in \mathbf{A} \cup \mathbf{S}$ is an attribute and $Edges$ contains (v^-, v^+, N) triples, where v^- and v^+ are values from $Dom(Atr)$ with $v^- < v^+$ and N is another trie node. A *leaf* node in a trie maintained by **TOSCA** is simply a set of tuples from the event KB, denoted $tuples(N)$. Tries have a unique *root* node.

A trie is *data correct* if for any leaf node N there is a unique path from the root $(Atr_1, Edges_1), \dots, (Atr_{k-1}, Edges_{k-1}), N$ such that for all $t \in tuples(N)$ and

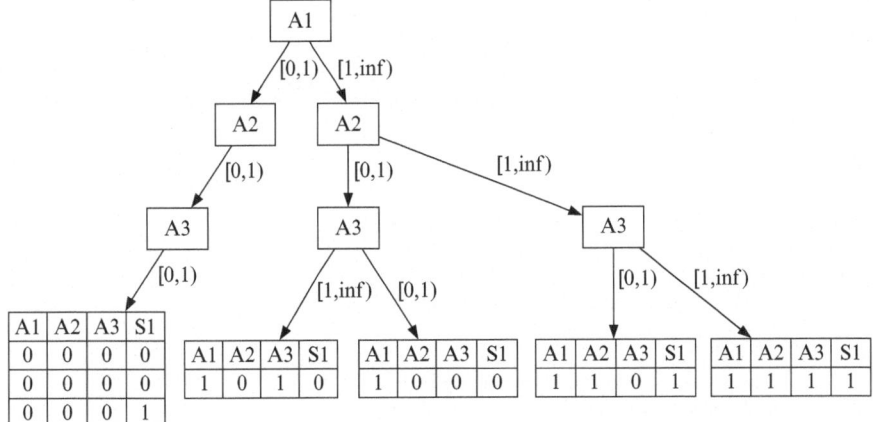

Fig. 3.1 On the left is an example trie data structure with Boolean action attributes $A1$, $A2$, and $A3$ and Boolean state attribute $S1$. The database the trie represents is depicted in Fig. 3.2

Database			
A1	A2	A3	S1
0	0	0	0
0	0	0	0
0	0	0	1
1	0	1	0
1	0	0	0
1	1	0	1
1	1	1	1

Fig. 3.2 The database indexed by the trie in Fig. 3.1

all i between 1 and $k - 1$, there is $(v^-, v^+, (Atr_{i+1}, Edges_{i+1})) \in Edges_i$ such that $v^- \le t(Atr_i) < v^+$. That is, the path to a leaf node determines which tuples are stored there. A trie is *construction correct* if for all sibling nodes (v_1^-, v_1^+, N_1) and (v_2^-, v_2^+, N_2), $v_1^- \ge v_2^+$ or $v_2^- \ge v_1^+$.

Trie creation is straightforward and is described in Algorithm 3. An essential decision for trie creation is the order in which the attributes appear along any given path. As such, we parameterize out the heuristic deciding the ordering; examining different heuristics is a topic of future work.

The TOSCA Algorithm. We now introduce the *Trie-enhanced Optimal State Change Attempt* (TOSCA) algorithm that uses tries to reduce the average case running time for computing optimal state change attempts. TOSCA is divided into the *base* and a *helper*, Algorithms 4 and 5, respectively. The following is an example of how TOSCA works.

Algorithm 3 createTrie(D,$\mathscr{A}tr$)

Return a trie for KB $\mathscr{K}$, using the attributes in the set $\mathscr{A}tr$. The function $choose$ is a heuristic that returns the "best" attribute to split on next

---

if $\mathscr{A}tr = \emptyset$ or $D = \emptyset$ **then**
   create leaf node N and set $tuples(N) = D$.
   return N.
end if
Let $A = choose(D, \mathscr{A}tr)$ // Pick an attribute to split on.
$Edges = \emptyset$ // The set of edges for the new node.
for $v \in Dom(A)$ **do**
   // Get the tuples in D with value v for attribute A.
   Let $D' = \sigma_{A=v}(D)$.
   if $D' \neq \emptyset$ **then**
      // Create a new trie containing the tuples in D'
      add $(D', createTrie(D', \mathscr{A}tr \setminus A))$ to $Edges$.
   end if
end for
return $(A, Edges)$.

Algorithm 4 TOSCA(Trie T, Goal G, Action tuple env, p)

Computes a state change attempt with minimal cost and probability of effectiveness at least p, using a trie rather than a bare database

---

1: Let $Dat1 = $ TOSCA-Helper(T, G, env).
2: Let $Dat2 = \emptyset$.
3: **for** $(SCA, f) \in Dat1$ **do**
4:    Let $nextF = pOccur(SCA|SCA) \cdot f$.
5:    **for** $(SCA', f) \in Dat1$ **do**
6:       **if** $SCA' \subsetneq SCA$ **then**
7:          $nextF = nextF + pOccur(SCA'|SCA) \cdot f'$
8:       **end if**
9:    **end for**
10:   Add (SCA, ef) with $(SCA, nextF)$ to $Dat2$.
11: **end for**
12: Remove any (SCA, ef) from $Dat2$ where $ef < p$.
13: **return** $argmin_{(SCA,ef) \in Dat2}(cost(SCA))$.

Example 3.5. In our example run of Algorithm 4, we use a simple database containing four tuples:

$$\{(A_1 = 1, E_1 = 1), (A_1 = 2, E_1 = 1), (A_1 = 3, E_1 = 0), (A_1 = 3, E_1 = 1)\},$$

and the trie T pictured in Fig. 3.3. We use the tuple $(A_1 = 0)$ as the action tuple env, the goal condition $E_1 = 1$, and the threshold 0.7 as p.

   The first step of Algorithm 4 is to create $Dat1$ via Algorithm 5, which recursively traverses the trie, beginning at node A. At node B, Algorithm 5 recognizes a leaf node and selects tuples from that node that satisfy the goal condition, iterating through them in turn beginning with $(A_1 = 1, E_1 = 1)$. The state

---

Algorithm 5 TOSCA-Helper(Trie T, Goal G, Action tuple *env*)

Returns a set of (SCA, v) pairs, where SCA is a state change attempt and v is $\varepsilon^*(\mathcal{K})(G, Sit = \gamma(SCA, env))$

1: **if** T is a leaf node **then**
2:     // The following is similar to Algorithm 2
3:     Let $Dat = \emptyset$
4:     **for** $t \in \sigma_G(tuples(T))$ **do**
5:         // Create SCA such that $\gamma(env, SCA) = t$
6:         Let $SCA = \{(A, env(A), t(A)) \mid t(A) \neq env(A)\}$
7:         If $(SCA, \cdot) \in Dat$ then continue to next t
8:         $f = \varepsilon_r^*(tuples(T))(t, G)$.
9:         Add (SCA, f) to Dat.
10:     **end for**
11:     **return** Dat.
12: **else**
13:     // Recursively call for all children of T.
14:     Let $(A, Edges) = T$.
15:     **return** $\bigcup_{(v-,v+,N) \in Edges}$ TOSCA-Helper(N, G, env)
16: **end if**

Fig. 3.3 The trie used in Example 3.5

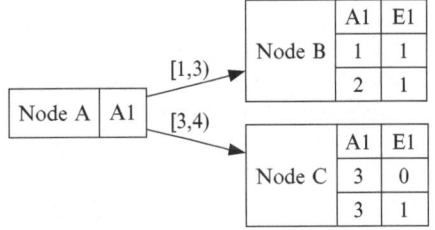

change attempt that changes the environment tuple $(A_1 = 0)$ to $(A_1 = 1)$ is $SCA = \{(A_1, 0, 1)\}$. The time saving step of the algorithm now occurs at line 8, where we run ε_r^* on the database $tuples(T)$ instead of the entire database (line 7 of Algorithm 2). Because there is only one tuple in $tuples(T)$ with $A_1 = 1$, and because that tuple also satisfies the goal condition, f is set to 1 and $(\{(A_1, 0, 1)\}, 1)$ is added to Dat. Similarly, $(\{(A_1, 0, 2)\}, 1)$ is added on the next tuple: $(A_1 = 2, S_1 = 1)$, finishing the call to node B.

The call to node C has slightly different results. The only member of $tuples(T)$ to satisfy the goal condition is $(A_1 = 3, S_1 = 1)$. Further, ε_r^* produces a result of $1/2$, as of the two tuples with value 3 for A_1, only one of them satisfies the condition that $S_1 = 1$. The returned set from this recursive call therefore contains only $(\{(A_1, 0, 3)\}, 1/2)$.

After merging all recursive calls, the set

$$\Big\{ \big(\{(A_1, 0, 3)\}, 1/2\big), \big(\{(A_1, 0, 2)\}, 1\big), \big(\{(A_1, 0, 1)\}, 1\big) \Big\}$$

is returned and labeled $Dat1$ by Algorithm 4. The next loop multiplies the second value of each member of $Dat1$ by the probability of the associated state change attempt occurring, which is provided by a user a priori and we will assume to be $3/4$ for all state change attempts. The inner loop then adds the probabilities associated with subsets of the state change attempt (of which there are none in this example). This results in the data structure $Dat2$ consisting of pairs $(SCA, pEff(env, S_1 = 1, SCA, \varepsilon_r^*))$, or

$$\left\{(\{(A_1, 0, 3)\}, 3/8), (\{(A_1, 0, 2)\}, 3/4), (\{(A_1, 0, 1)\}, 3/4)\right\}.$$

At this point, those members of $Dat2$ with too low a probability of effectiveness are eliminated (only $(\{A_1, 0, 3\}, 3/8)$), and the SCA with lowest cost is returned. ■

Proposition 3.3. *Algorithm 4 computes state change attempt SCA satisfying the following properties:*

1. *$pEff(env, G, SCA, \varepsilon_r^*(\mathcal{K})) \geq p$, and*
2. *There is no other applicable state change attempt SCA' such that $cost(SCA') < cost(SCA)$ and $pEff(env, G, SCA', \varepsilon_r^*(\mathcal{K})) \geq p$.*

Proof. To show this, it suffices to show that in line 15 for all $(SCA, ef) \in Dat2$, $v = pEff(env, G, SCA, \varepsilon_r^*)$. Consider any $(SCA, ef) \in Dat2$ on that line, and note that due to the loop starting in line 3, we have:

$$ef = \sum_{(SCA', f') \in Dat1, SCA' \subseteq SCA} f' \cdot pOccur(SCA' \mid SCA).$$

Since for all $(SCA', f') \in Dat1$, $f' = \varepsilon_r^*(tuple(T))(\gamma(env, SCA), G)$, (from TOSCA-Helper) where $tuples(T)$ is the set of all tuples satisfying $\gamma(env, SCA)$, this suffices to show that $ef = pEff(SCA, G, env, \varepsilon_r^*)$.

Finally, lines 12 and 13 guarantee that the only SCAs returned are those that have probability of effectiveness at least p and cost at most k. □

The worse case time complexity of Algorithm 4 is $O(|\mathcal{K}|^2)$. However, the complexity of Algorithm 5 is $O(|\mathcal{K}| \cdot k)$, where k is the size of the largest leaf node in trie T. While k is bounded by $|\mathcal{K}|$, it is usually much smaller: on the order of $|\mathcal{K}|/2^h$ for a trie of height h. We expect $Dat1$ to have size $O(|\mathcal{K}|)$, as it will be the same as $Dat1$ in line 10 of Algorithm 2; however, it was produced by at most $2 \cdot |\mathcal{K}|/k$ recursive calls to Algorithm 5 (there are at most $2 \cdot |\mathcal{K}|/k$ nodes in the trie). When given a leaf node, Algorithm 5 takes time in $O(k^2)$. Thus, the running time of Algorithm 5 is in $O(|\mathcal{K}| \cdot k)$. The loop in line 3 then runs in time in $O(|\mathcal{K}|^2)$ (it is the same loop as in Algorithm 2), resulting in an overall run time in $O(|\mathcal{K}|^2)$. However, we will see that in practice, substantial speedup is gleaned from using the $O(|\mathcal{K}| \cdot k)$ Algorithm 5 rather than the basic $O(|\mathcal{K}|^2)$ approach.

We will conclude this chapter with a brief discussion of two proposals for further improvements that can be made to the trie-based approach.

3.4.1 Reducing Trie Size by Bucketing Values

The fact that each distinct value of each attribute produces a new edge in the trie means that the structure will grow in size quite rapidly. To ameliorate this problem, we propose dividing the domain of each attribute into *buckets*. This division need not be at even intervals, since we might prefer to define the buckets so that they are all more or less of the same size given the KB. There are two important ways in which this modification impacts our structure: first, we can decide what the branching factor is at each internal node, and second, we can prune any branches that correspond to values that do not satisfy condition C in Algorithm 5. Another thing to note is that leaf nodes will in general contain more tuples than before.

As mentioned, the way in which search is performed in the trie must also be adapted. We can prune any branches that are guaranteed to yield a value of zero when the selection in line 8 is performed. For instance, if C is the condition *"Graduation% $\geq$ 85 and Reading Proficiency $\geq$ 75"*, then the recursive calls in line 15 of Algorithm 5 should disregard all buckets that have no possibility of satisfying either condition, for instance *Graduation%* $\in [0, 10]$, $[11, 20]$, etc.

3.4.2 Annotated Tries

Similar to the ideas described on bucketing values described above, we can annotate the internal nodes of the trie so that search can be made more efficient. The annotations we propose consist simply of the number of tuples that exist in the database for each value (or each bucket) of the corresponding attribute that are below that node in the trie. This can be done easily during the trie creation process, with an overhead of one additional selection query for each possible value (or bucket) of each attribute.

The use of these annotations during search is also similar to the use of buckets. That is, each recursive call in line 15 will be performed only if the value (or bucket) in question has the possibility of satisfying condition C, and its annotation is not zero. Note that this is a refinement of the way in which search is performed when only bucketing is used.

References

1. Edward Fredkin. Trie memory. *Communications of the ACM*, 3(9):490–499, 1960.
2. Tom M. Mitchell. *Machine Learning*. McGraw-Hill, New York, 1997.
3. R. Rojas. *Neural Networks: A Systematic Introduction*. Springer, 1996.

Chapter 4
A Comparison with Planning Under Uncertainty

In order to investigate how our approach to solving the proposed class of problems relates to traditional approaches such as planning under uncertainty, in this chapter we will propose and discuss a mapping between an instance of an OSCA problem and an instance of a *Markov Decision Process*. The ultimate goal is to show that optimal state change attempt problems can indeed be solved by applying techniques from the planning under uncertainty literature, but this approach will be ultimately impractical.

First of all, let us recall the elements required to describe an instance of an OSCA problem:

1. Set of *action attributes*: corresponds to the actionable attributes that we can potentially *act upon* in order to change their values.
2. Set of *state attributes*: used to describe the *situations* of the environment that we cannot directly change, including the outcome attributes that we may want to influence.
3. A *cost function* for state change attempts: describes the cost of changing the values of the action attributes.
4. An *effect estimator*: describes the conditional probability that a given goal holds given an assignment of values to the action attributes.
5. *Conditional probabilities* for probability of occurrence of SCAs: describes the probability that a certain state change attempt is successful given that another state change was attempted.
6. A *goal* specified over the values of a subset of the state attributes: describes the state of affairs that the user desires to accomplish.

The objective is to compute an optimal state change attempt with respect to cost and/or probability of effectiveness given the goal; such an SCA intuitively represents a policy that can be applied with the objective of reaching that specific goal. On the other hand, in order to describe an instance of an MDP we require:

1. A finite set S of environment *states*.
2. A finite set A of *actions*.

A. Parker et al., *Data-driven Generation of Policies*, SpringerBriefs in Computer Science, DOI 10.1007/978-1-4939-0274-3_4, © The Author(s) 2014

3. A *transition function* $T : S \times A \to \Pi(S)$ specifying the probability of arriving at every possible state given that a certain action is taken in a given state.
4. A *reward function* $R : S \times A \to \mathbb{R}$ specifying the expected immediate reward gained by taking an action in a state.

The objective is to compute a policy $\pi : S \to A$, specifying what action should be taken in each state, that is optimal with respect to the expected utility obtained from executing it.

4.1 Obtaining an MDP from the Specification of an OSCA Problem

We will now propose how, given an instance of an OSCA problem as described above, we can obtain the specification of a corresponding MDP in such a way that optimal policies for this MDP correspond to solutions to the original OSCA problem.

- *State Space*: The set S_{MDP} of MDP states corresponds to the set of all possible tuples $(v_1, \ldots, v_m) \in dom(S_1) \times \ldots \times dom(S_m)$, where $\bigcup_{i=1}^{m} S_i = \mathbf{S}$, the set of all state attributes.
- *Actions*: The set A_{MDP} of possible actions in the MDP domain corresponds to the set of all possible state change attempts. Without considering the fact that not all SCAs will be applicable in every state, we can think of the set of actions as containing any subset of h action attributes, each of which can be attempted to be changed to any other possible value in its domain.
- *Transition Function*: The (conditional) probabilities of occurrence can be used to define the transition function T for the MDP, since it is clear what the effect of a change attempt is when it is successful. Formally, let $s, s' \in S_{MDP}$ and $a \in A_{MDP}$; if $s = (u_1, \ldots, u_m)$, $s' = (u'_1, \ldots, u'_m)$, and $a = ((A^1_{i_1}, vf_1, vt_1), \ldots, (A^h_{i_h}, vf_h, vt_h))$ for $i_1, \ldots, i_h \in \{1, \ldots, |\mathbf{A}|\}$, we define:

$$
T(s, a, s') = \begin{cases} 0 & \text{if } a \text{ is not applicable in } s, \\ pEff(s, G_{s'}, a, \varepsilon) & \text{otherwise.} \end{cases} \tag{4.1}
$$

where $G_{s'}$ denotes the condition that imposes the values in s' on the state attributes as the goal. However, if the OSCA problem requires the *lowest cost* solution only (see Sect. 2.4), we simply define T to be as follows. Let $S_a \subseteq S_{MDP}$ be the set of all states s' for which $pEff(s, G_{s'}, a, \varepsilon) \neq 0$:

$$
T(s, a, s') = \begin{cases} 0 & \text{if } a \text{ is not applicable in } s, \\ \frac{1}{|S_a|} & \text{otherwise.} \end{cases} \tag{4.2}
$$

- *Reward Function*: The reward function of the MDP, which describes the reward directly obtained from performing action $a \in A$ in state $s \in S$, is defined based on two aspects: (1) the probability that a state satisfying the goal is reached by taking action a in state s (this will depend on the *effect estimator* being used), and (2) the cost of the change attempt associated with action a. It should be noted that, as in the case of the transition function above, the specific problem to be solved (lowest cost, highest probability, etc.), will directly influence the way in which the corresponding reward function is defined (e.g., for highest probability problems, cost is ignored). Let G be the goal corresponding to the OSCA problem instance and, as above, let $s \in S_{MDP}$ and $a \in A_{MDP}$ such that $s = (u_1, \ldots, u_m)$ and $a = \left((A_{i_1}^1, vf_1, vt_1), \ldots, (A_{i_h}^h, vf_h, vt_h)\right)$ for $i_1, \ldots, i_h \in \{1, \ldots, |\mathbf{A}|\}$:

$$R(s,a) = \begin{cases} 0 & \text{if } a \text{ is not applicable in } s, \\ pEff(s, G, a, \varepsilon) & \text{otherwise.} \end{cases} \tag{4.3}$$

Similarly, for lowest cost, we have:

$$R(s,a) = \begin{cases} 0 & \text{if } a \text{ is not applicable in } s, \\ pEff(s, G, a, \varepsilon) * \frac{1}{cost(a)} & \text{otherwise.} \end{cases} \tag{4.4}$$

As can be seen by the above mapping, the *key point in which our problem differs* from planning problems is that SCAs involve executing actions in parallel which, among other things, means that the number of possible simple SCAs that can be considered in a given state is *very large*. This makes planning approaches infeasible since their computational cost is intimately tied to the number of possible actions in the domain (generally assumed to be fixed at a relatively small number). In the case of MDPs, even though state aggregation techniques have been investigated to keep the number of states being considered manageable [1, 3, 6], similar techniques for *action aggregation* have not been developed.

To conclude this comparison with planning under uncertainty using MDPs, we present the following results. The first states that given an instance of OSCA, the proposed translation into an MDP is such that an optimal policy under Maximum Expected Utility (MEU) for such an MDP expresses a solution for the original instance. Note, however, that such a policy is actually a *fully contingent* plan in that it prescribes an action for every possible state; this means that the state change attempt prescribed in each state is chosen taking into account what would happen if the goal is not immediately reached, and thus states that have a better utility computed in this manner are preferred. A fair comparison with the approach taken in this work would be to have our *OSCA* algorithms given above iterate until the goal is satisfied.

Proposition 4.1. *Let $O = (\mathbf{A}, \mathbf{S}, cost, \varepsilon, pOccur, G)$ be a specification of an OSCA problem and $M = (S_{MDP}, A_{MDP}, T, R)$ be its corresponding translation into an MDP. If π is a policy for M that is optimal with respect to the MEU criterion, then* *for any state $s \in S_{MDP}$, $\pi(s)$ yields a state change attempt that is a solution for O for the values of the state attributes described by s.*

Proof (sketch). Assume that the OSCA instance given is a *highest probability* instance (the lowest cost case is analogous). By hypothesis we have that π is MEU-optimal, which means that

$$\pi(s) = \arg\max_a \left(R(s,a) + \max_{a'} \left(\sum_{s' \in S} T(s,a,s') \cdot Q(s',a') \right) \right) \qquad (4.5)$$

where Q is the action utility function defined as usual. Now, suppose towards a contradiction that there exists a state s such that the state change attempt corresponding to $a = \pi(s)$ is sub-optimal, i.e., there exists another state change attempt $a' = \pi'(s)$ that has a higher probability of effectiveness in reaching the goal; formally, $pEff(s, G, \pi'(s), \varepsilon) > pEff(s, G, \pi(s), \varepsilon)$. Since both the reward and transition functions are defined in terms of probability of effectiveness, this directly implies that

$$\left(R(s,a') + \max_b \left(\sum_{s' \in S} T(s,a',s') \cdot Q(s',b) \right) \right) >$$

$$\left(R(s,a) + \max_b \left(\sum_{s' \in S} T(s,a,s') \cdot Q(s',b) \right) \right)$$

However, this contradicts Eq. 4.5 above since $\pi(s)$ was selected as the state change attempt that maximizes this sum. The contradiction stemmed from the assumption that $\pi(s)$ is sub-optimal; therefore, we can conclude that $\pi(s)$ corresponds to a solution to O.   □

Second, we analyze the computational cost of taking this approach. Since there exists in the literature a large variety of algorithms for solving MDPs, we will only analyze the size of the MDP resulting from the translation of an instance of OSCA.

Proposition 4.2. *Let $O = (\mathbf{A}, \mathbf{S}, cost, \varepsilon, pOccur, G)$ be a specification of an OSCA problem and $M = (S_{MDP}, A_{MDP}, T, R)$ be its corresponding translation into an MDP. Then, we have that:*

$$|S_{MDP}| = \prod_{i=1}^{m} |dom(S_i)|,$$

where $S_i \in \mathbf{S}$, and:

$$|A_{MDP}| \leq \sum_{h=1}^{n} \left(A_h^n * |V|^h \right),$$

where $n = |\mathbf{A}|$, and $|V|$ is the maximum number of possible values for an action attribute.

Proof. For the size of the state space, simply recall that the MDP's state space corresponds to the set of all possible tuples $(v_1, \ldots, v_m) \in dom(S_1) \times \ldots \times dom(S_m)$, where $\bigcup_{i=1}^{m} S_i = \mathbf{S}$.

In order to prove the upper bound on the number of actions, suppose that all SCAs will be applicable in every state. Then, we have $|A_{MDP}| = \sum_{h=1}^{n} \left(A_h^n * |V|^h \right)$, where n is the number of actions in the action schema $\mathbf{A}$, and V is the set of possible values for the action attributes (we assume $|dom(A_i)| = |dom(A_j)|$ for all $A_i, A_j \in \mathbf{A}$ for the purpose of this analysis). Basically, this formula states that any combination of h action attributes can be chosen, and each can be attempted to be changed to any other possible value in its domain. $\square$

Consider that, for instance, the well-known Value Iteration algorithm [2, 5] iterates over the entire state space a number of times that is polynomial in $|S|$, $|A|$, β, and B, where B is an upper bound on the number of bits that are needed to represent any numerator or denominator of β [4]. Now, each iteration takes time in $O(|A| \cdot |S|^2)$, meaning that only for very small instances will MDPs of the size expressed in Proposition 4.2 be feasible.

References

1. Craig Boutilier, Richard Dearden, and Mosiés Goldszmidt. Stochastic dynamic programming with factored representations. *Artificial Intelligence*, 121(1–2):49–107, 2000.
2. Richard Bellman. A markovian decision process. *Journal of Mathematics and Mechanics*, 6, 1957.
3. Thomas Dean, Robert Givan, and Sonia Leach. Model reduction techniques for computing approximately optimal solutions for Markov decision processes. In Dan Geiger and Prakash Pundalik Shenoy, editors, *Proceedings of the 13th Conference on Uncertainty in Artificial Intelligence (UAI-97)*, pages 124–131, San Francisco, August 1–3 1997. Morgan Kaufmann Publishers.
4. Michael Lederman Littman. *Algorithms for Sequential Decision Making*. PhD thesis, Department of Computer Science, Brown University, Providence, RI, February 1996.
5. M. L. Puterman. *Markov decision processes: Discrete Stochastic Dynamic Programming*. John Wiley and Sons, Inc., New York, 1994.
6. John Tsitsiklis and Benjamin van Roy. Feature-based methods for large scale dynamic programming. *Machine Learning*, 22(1/2/3):59–94, 1996.

Chapter 5
Experimental Evaluation

In this chapter, we will describe a set of empirical results obtained from a prototype implementation of limitedSCASet (Algorithm 1), DSEE_OSCA (Algorithm 2) and TOSCA (Algorithm 4). limitedSCASet was implemented in a modular way so that one could change the effect estimator, while the other algorithms were implemented assuming the data ratio effect estimator (Definition 3.4). We did an experimental analysis to determine the outcomes of four major questions:

1. Which kind of effect estimator gives the most accurate results?
2. Which techniques provide the best running time with large amounts of data?
3. Which techniques provide the best running time as the number of attributes and their domain size increases?
4. Which techniques provide the best running time with real-world data?

In the following, we will present our findings for each case.

5.1 Question 1: Which Effect Estimator Gives the Most Accurate Results?

To address this question, we used the Weka framework's implementation of several machine learning algorithms [5]. These include the AODE algorithm, which creates Bayes nets [4]; the IBk algorithm, which uses K-nearest neighbor clustering [1]; and the C4.5 algorithm, which uses entropy minimization techniques to create decision trees that can be used for classification [2]. We used these algorithms to implement respective learned effect estimators (Definition 3.2) and combined those effect estimators with Algorithm 1 to get several different methods for determining the optimal state change attempt. We compared these to the data ratio effect estimator.

To generate the data in this experiment, we produced k tuples with four action attributes and three state attributes. Each tuple's value for the action attributes was chosen randomly from $[0, 1]$. To generate the values for the state attribute, we

A. Parker et al., *Data-driven Generation of Policies*, SpringerBriefs in Computer Science, DOI 10.1007/978-1-4939-0274-3_5, © The Author(s) 2014

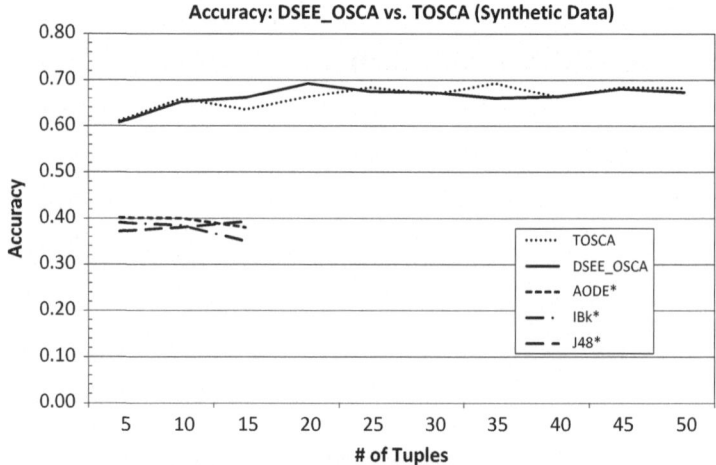

Fig. 5.1 A comparison of the accuracy of all algorithms being evaluated, over synthetic data. The learning effect estimators clearly are outperformed by the data ratio effect estimator; Algorithms marked with * took longer than 2 days to complete for inputs of 20 or more tuples

generated random boolean formulas over the action attributes consisting of the operators $<, >, =, \neq$, and $\wedge$. We allowed at most three "$\wedge$" connectives in each formula. In a given tuple, each state attribute value is set to 1 if its associated formula is satisfied by the action attributes in that tuple, and set to 0 otherwise. Because we have the formula defining the state attributes, we can easily check the accuracy of the state change attempts returned by each algorithm. To do this, we apply the state change attempt and determine the state attribute values. The accuracy of a given algorithm will be the fraction of the time the resulting values for the state attributes satisfy the goal condition.

For this experiment, the goal condition is generated randomly as above, the cost of each simple state change is 1, and the probability of occurrence is set to 1 (all state change attempts always occur as expected).

The results of these experiments are shown in Figs. 5.1, and 5.2. In these figures we notice two things. First, we notice that the data ratio effect estimator (used in **DSEE_OSCA** and **TOSCA** in the graphs) runs substantially faster than the learning effect estimators. This is due to the algorithms needed for both estimators. Since the data ratio effect estimator assumes that anything not occurring in the database is a negative instance, it needs to consider substantially fewer possibilities than the learning effect estimators. The learning effect estimators, on the other hand, allow for any possible tuple to satisfy the goal condition and do not need to limit themselves to those tuples already in the database. This generality incurs substantial computational costs. Since they consider every combination of all the attribute values in the database, they are only able to finish computation when there are relatively few tuples. Normally one would hope that this generality would nonetheless result in an increase in accuracy—since learning effect estimators

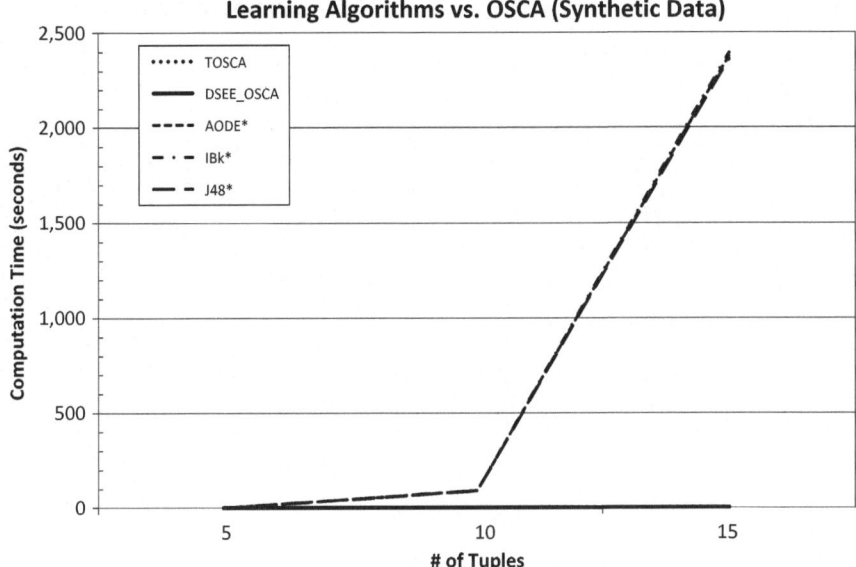

Fig. 5.2 A comparison of average running time over at least 60 runs for the data selection OSCA approach (DSEE_OSCA and TOSCA) and learning algorithms (AODE, IBk, and J48) over synthetic data. The learning algorithms took longer than two days to compute the optimal SCA for 20 or more tuples; DSEE_OSCA and TOSCA's running times correspond to the two curves that are touching the x-axis

consider more possible solutions than data ratio effect estimators they should be able to find a better solution. However, in these experiments that is not the case: by restricting itself to only those possible solutions currently in the database, data ratio effect estimators actually increase their accuracy over the various learning effect estimators.

Finally, Fig. 5.3 shows the comparison of the accuracy of DSEE_OSCA versus that of TOSCA. As expected, since both algorithms make use of the data selection effect estimator, the accuracy of both algorithms is comparable, as shown in the figure.

5.2 Question 2: Which Techniques Scale Best?

We use the same experimental setup as in Question 1 (Sect. 5.1) to see how DSEE_OSCA and TOSCA scale when presented with larger amounts of data. Since the algorithms using the learning effect estimators could not scale past 20 tuples, they are not included in this experiment.

In these experiments, we provided the algorithms with 1–15,000 tuples. The results are shown in Fig. 5.4, and they show that TOSCA performs better than DSEE_OSCA as the database increases in size. We should note that TOSCA does

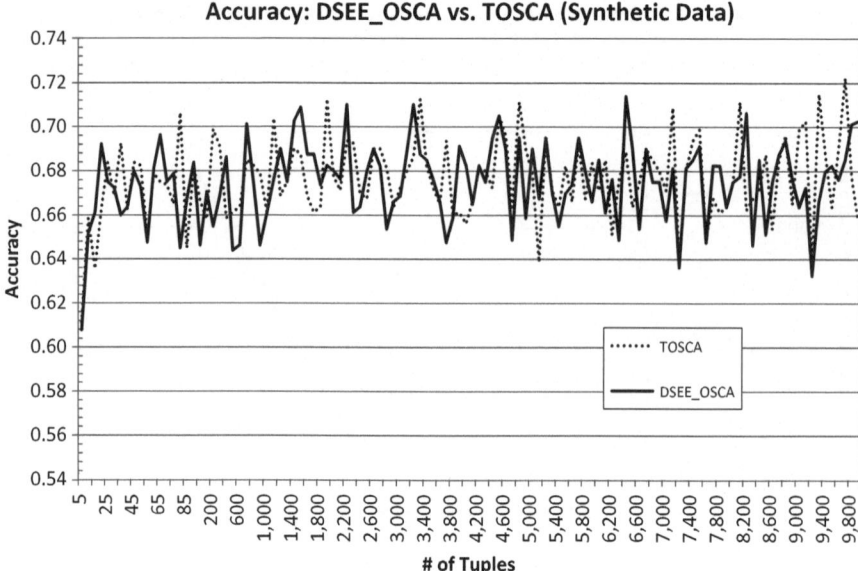

Fig. 5.3 A comparison of the accuracy of DSEE_OSCA to that of TOSCA, over synthetic data

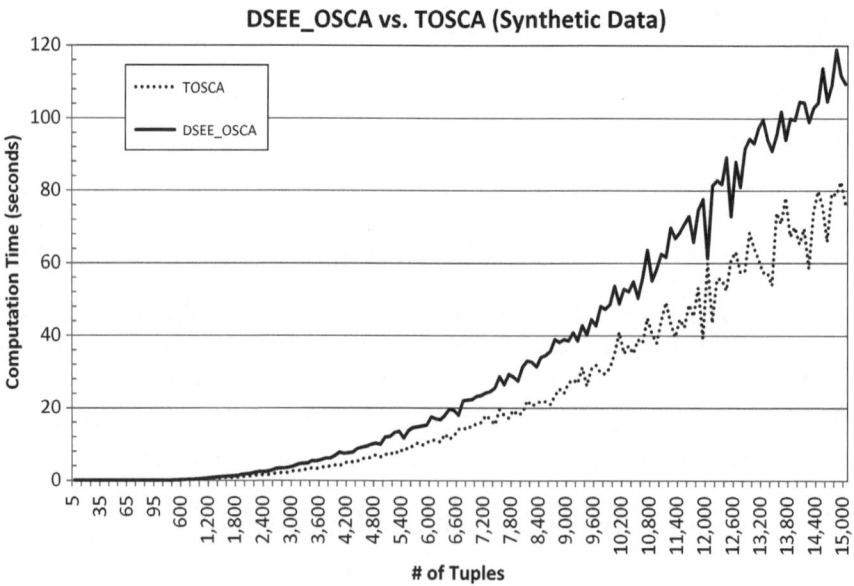

Fig. 5.4 A comparison of average running time over at least 60 runs for DSEE_OSCA and TOSCA over synthetic data

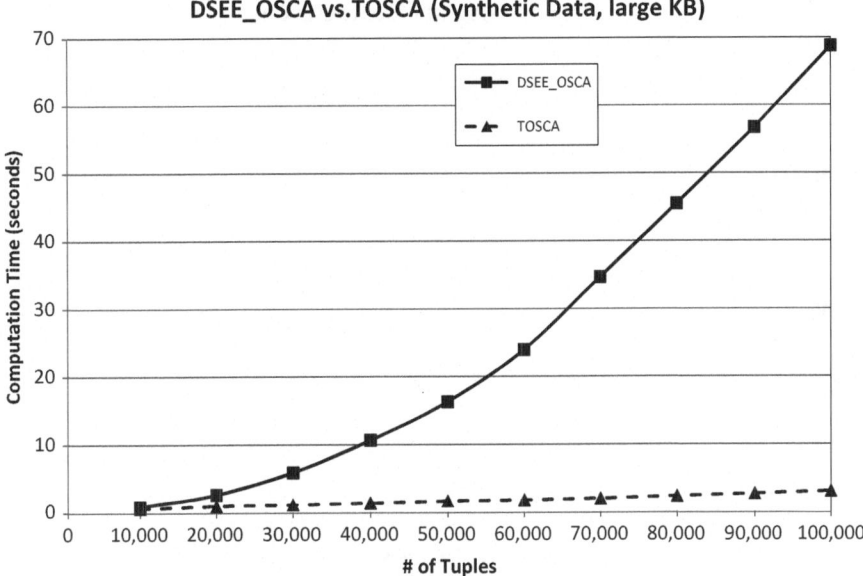

Fig. 5.5 Performance of the naive algorithm DSEE_OSCA (Algorithm 2) versus the TOSCA (Algorithm 4) in artificial data experiments with large amounts of data

have a pre-computation step whose running time has been left out of these figures. However, the time needed to compute the trie is several orders of magnitude smaller than the running time of TOSCA: for instance, it takes only 91 ms to construct the trie with 10,000 tuples.

Another configuration of synthetic data generation: For the following set of experiments, we used a schema with four action attributes and one state attribute, each with domain containing the integers 0–99 (inclusive). For each trial, a new database was generated where each tuple's attributes were chosen uniformly at random from the associated domain.

For each action attribute $A_i \in \mathbf{A}$, we randomly assigned a basic probability of occurrence $P(A_i)$ (real-world applications would infer this from historical data). For state change attempts SCA and $SCA' \supseteq SCA$, $pOccur$ was estimated as follows:

$$pOccur(SCA \mid SCA') = \prod_{(A_i, vf, vt) \in SCA} \frac{P(S_i)}{cost((A_i, vf, vt))}$$

For our experiments we computed the cost of a simple state change attempt (A_i, vf, vt) as the simple distance between vf and vt; that is, $cost((A_i, vf, vt)) = |vt - vf|$.

The results are shown in Figs. 5.5 and 5.6. The former shows a comparison between the running times of the two algorithms for KBs up to 100k tuples, and clearly shows that TOSCA outperforms DSEE_OSCA. The latter only shows how

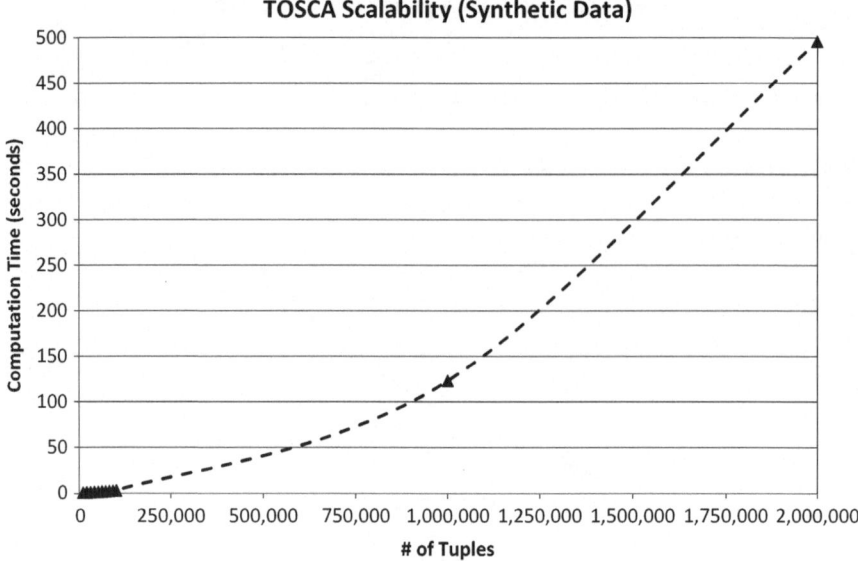

Fig. 5.6 Performance of TOSCA (Algorithm 4) including the time taken by the procedure creating the trie data structure being used. These runs were carried out over artificial data

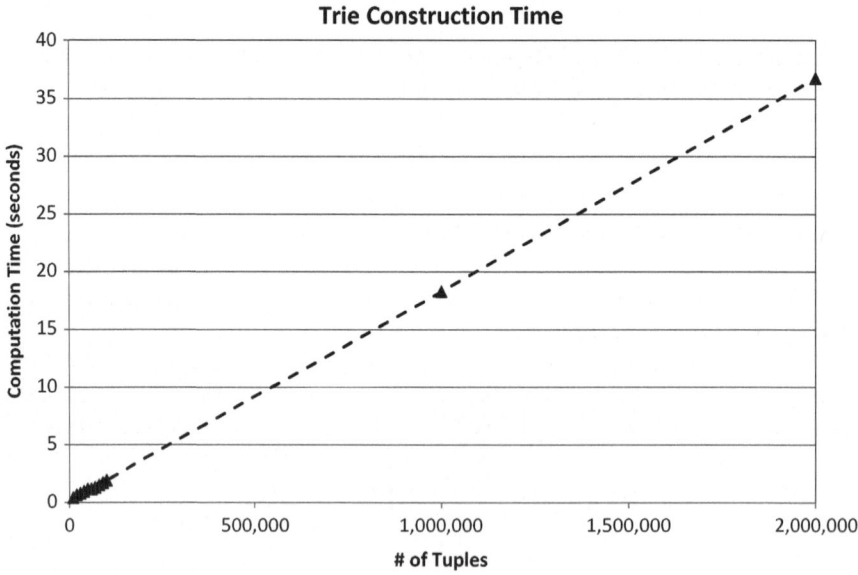

Fig. 5.7 A plot of the time taken by the procedure creating the trie data structure being used in the TOSCA algorithm for the same runs reported in Fig. 5.6 (artificial data)

TOSCA performs with KBs of up to two million tuples; note that the times reported for TOSCA include the time needed to compute the trie (though presumably, such an index would be pre-computed in practice). Finally, the trie construction times for the same runs reported in Fig. 5.6 are shown in Fig. 5.7.

5.3   Question 3: Which Techniques Provide the Best Running Time as the Number of Attributes and Their Domain Size Increases?

In Fig. 5.8 we see how DSEE_OSCA and TOSCA scale as the number of attributes increases in a database with 8,000 tuples. None of the learning algorithm effect estimators are depicted because they do not scale to such a large database. This graph shows TOSCA outperforming DSEE_OSCA. This graph is important because the trie in TOSCA should lose efficiency as the number of attributes increases (the trie's depth equals the number of attributes). However, this graph shows that decrease in the trie's efficiency does not affect the ability of the trie to offer TOSCA a speedup over DSEE_OSCA.

The effect of varying the size of the domains of the action attributes is shown in Fig. 5.9, where the size of the event KB was fixed at 8,000 tuples, each with four action attributes and three state attributes. The plot shows the DSEE_OSCA and

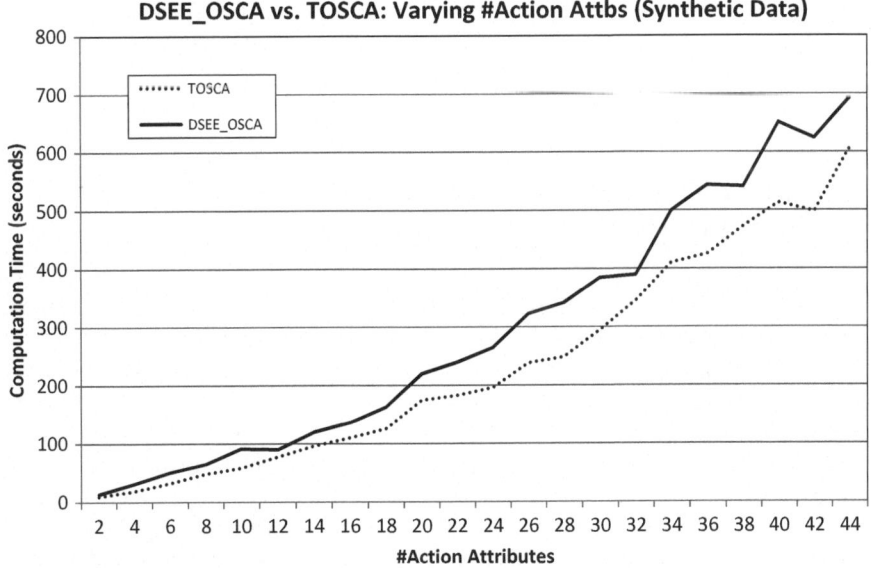

Fig. 5.8 The running times of DSEE_OSCA and TOSCA (over synthetic data) as the number of action attributes increases and number of tuples is fixed at 8,000

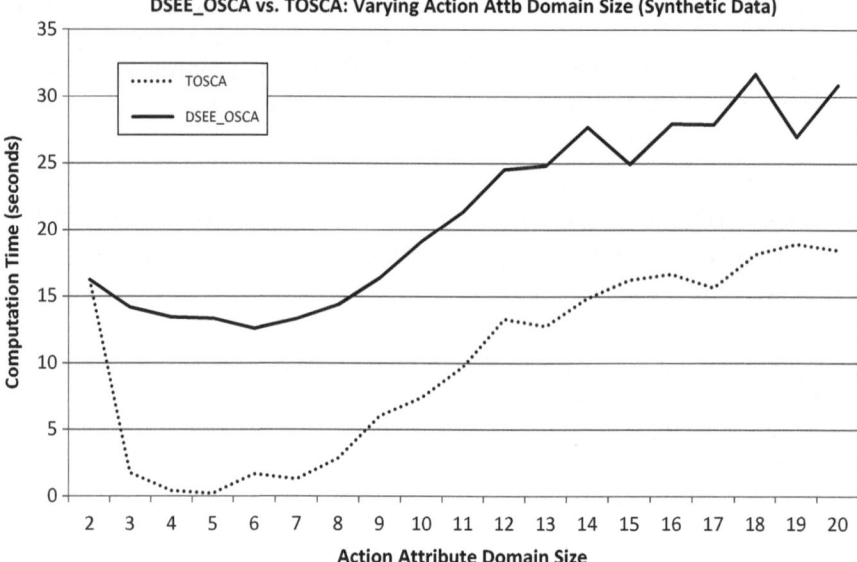

Fig. 5.9 Performance of the naive algorithm DSEE_OSCA (Algorithm 2) versus the TOSCA (Algorithm 4) in synthetic data experiments when the size of the domain of action attributes is varied. The number of tuples was fixed at 8,000, action attributes at 4, and state attributes at 3

TOSCA algorithms performing equally well for domains of size 2, which makes sense since in this case the trie cannot be leveraged. For larger domain sizes, we see a fairly constant difference in favor of TOSCA, which reflects the expected speedup that does not vary since the size of the KB is fixed.

5.4   Question 4: Which Algorithms Perform Best with Real-World Data?

We used the U.S. School Dataset [3], which includes data on school performance, budget, and related variables for all schools in a given state, as described in Chap. 1 (Example 1.1). We chose the state of Arizona for our tests. The variables can naturally be divided into action and state attributes; to keep our experiments manageable, we chose a subset of about 50 of the most important attributes, preferring the general variables over the specific ones. We then varied the number of attributes of the chosen 50 that would be considered action attributes, including sets of 6, 12, and 24 action attributes. Again, since the algorithms using the learning effect estimators could not scale past 20 tuples, they are not included in this experiment. In each run, we took a random subset of the appropriate size from one state's data in the U.S. school data database and then generated a random

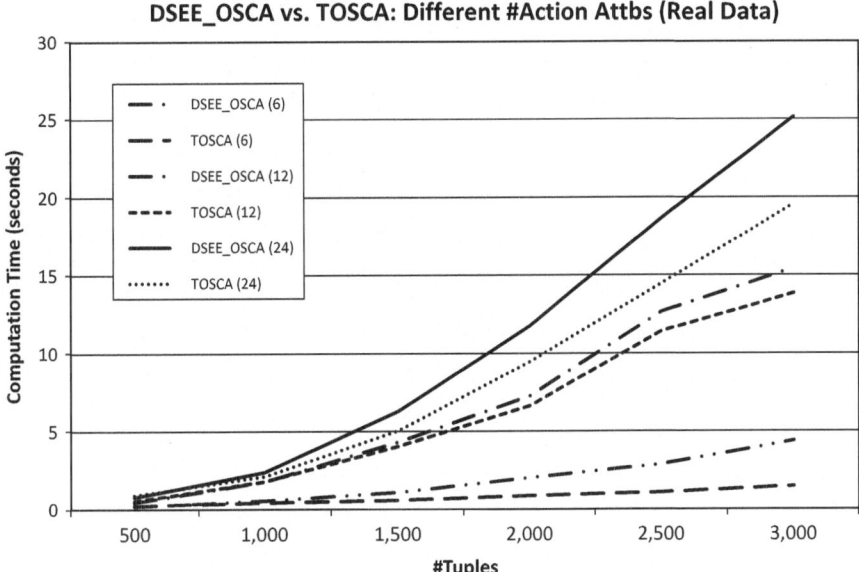

Fig. 5.10 Performance of the naive algorithm DSEE_OSCA (Algorithm 2) versus the TOSCA (Algorithm 4) in real data experiments. This figure plots three comparisons in one, for 6, 12, and 24 action attributes each

goal condition based on the domains of the provided action variables. We then ran both DSEE_OSCA and TOSCA, keeping track of the running times, including the running time of building the trie for TOSCA. The results of these experiments are shown in Fig. 5.10 and clearly show that TOSCA is faster than DSEE_OSCA for any given number of action attributes.

References

1. D. Aha and D. Kibler. Instance-based learning algorithms. *Machine Learning*, 6:37–66, 1991.
2. Ross Quinlan. *C4.5: Programs for Machine Learning*. Morgan Kaufmann Publishers, San Mateo, CA, 1993.
3. State Education Data Center. http://www.schooldatadirect.org, 2008.
4. G. Webb, J. Boughton, and Z. Wang. Not so naive bayes: Aggregating one-dependence estimators. *Machine Learning*, 58(1):5–24, 2005.
5. Ian H. Witten and Eibe Frank. *Data Mining: Practical machine learning tools and techniques, 2nd Edition*. Morgan Kaufmann, San Francisco, 2005.

Chapter 6
Conclusions

The AI planning literature contains decades of substantial work on discovering sequences of actions that lead to a given outcome that, similar to this work, is often specified as a goal condition. In this work, we have described an approach to solve problems that at first seems quite similar to those tackled by AI planning; however the main characteristic of the problems of interest are that important assumptions made in AI planning approaches cannot be made in this case. There are, however, significant assumptions made in these works that cannot always be made, such as:

1. The number of actions available to solve the problem is assumed to be relatively small,
2. The causal relationships within the model describing the environment are well understood (e.g., if a block is picked up and put on top of another block, then the rest of the blocks remain unchanged, and the new pile continues to exist until some further action changes it).
3. The effects of actions taken in the environment are well understood (e.g., in the example above, picking up the block only has the effect of the block no longer being on the table).

In this work, we have described an approach to solve problems that at first seems quite similar to those tackled by AI planning; however the main characteristic of the problems of interest are that the above assumptions cannot be made: effects of actions are not clearly understood, nor are the causal relationships between the values of the parameters in the system, and there are a huge number of actions to choose from. Our proposal was therefore to solve these problems in a data-driven manner, where these poorly understood effects and relationships can be gleaned from the data instead of assuming that they are packaged with the input.

In Chap. 1 we began by introducing this problem and the concept of event databases, where attributes are assumed to be either of type *action* or *state*; the former are those that can directly be manipulated (albeit at a certain cost and with certain probability of success), while the latter constitute those that can only be influenced indirectly. Two examples of this kind of data were introduced: school performance Fig. 1.1 and city government Fig. 1.2. In Chap. 2, we formally

A. Parker et al., *Data-driven Generation of Policies*, SpringerBriefs in Computer Science, DOI 10.1007/978-1-4939-0274-3_6, © The Author(s) 2014

introduced the different pieces of this problem: (optimal) state change attempts, effect estimators, cost functions, and probability of effectiveness, illustrating each of them on the datasets from the running examples. We also showed that determining optimal state change attempts is not an easy problem, proving that most interesting versions of the optimization task belong to complexity classes widely believed to be intractable. In Chap. 3, we began by taking a closer look at the different kinds of effect estimators that can be defined, and focused on the *data selection* effect estimator due to its computational properties. Also in this chapter, we introduced the TOSCA algorithm, an approach that combines data selection effect estimators with trie data structures to obtain optimal state change attempts more efficiently. In Chap. 4, we explored how optimal state change attempts can be computed by applying Markov Decision Processes, one of the most widely used models in planning under uncertainty, and showed that, even though it is theoretically possible, the size of the resulting problem prohibits its application in practice. Finally, in Chap. 5 we presented empirical results obtained from running implementations of our algorithms both on synthetic and real-world datasets, showing that the TOSCA algorithm can be applied to event databases with hundreds of thousands, and even millions. of tuples.

Index

A. Parker et al., *Data-driven Generation of Policies*, SpringerBriefs in Computer Science,
DOI 10.1007/978-1-4939-0274-3, © The Author(s) 2014